Monitoring HumanTissues for Toxic Substances

Committee on National Monitoring
of Human Tissues

Board on Environmental Studies and Toxicology

Commission on Life Sciences

NATIONAL ACADEMY PRESS
Washington, D.C. 1991

NATIONAL ACADEMY PRESS   2101 Constitution Ave., N.W.   Washington, D.C. 20418

Library of Congress Catalog Card No. 91-61252
International Standard Book Number 0-309-04437-5

Additional copies of this report are available from the National Academy Press, 2101 Constitu- tion Avenue, N.W., Washington, D.C. 20418

S289

Printed in the United States of America

Committee on National Monitoring of Human Tissues

JOHN C. BAILAR, III *(Chairman)*, McGill University School of Medicine, Montreal

DAVID GAYLOR, U.S. Food & Drug Administration, National Center for Toxicologic Research, Jefferson, Arkansas

WILLIAM GRIZZLE, University of Alabama at Birmingham

THOMAS GRUMBLY, Clean Sites, Inc., Alexandria, Virginia

DAVID KALMAN, University of Washington, Seattle

KATHRYN MAHAFFEY, National Institute of Environmental Health Sciences and University of Cincinnati Medical School

H. B. MATTHEWS, National Institute of Environmental Health Sciences, Research Triangle Park

FREDERICA PERERA, Columbia University, New York

JOSEPH WAKSBERG, WESTAT, Rockville, Maryland

Staff

LEE R. PAULSON, Project Director

CAROLYN FULCO, Staff Officer (until June 1990)

NORMAN GROSSBLATT, Editor

BERNIDEAN WILLIAMS, Information Specialist

SHELLEY NURSE, Senior Project Assistant

Sponsor: U.S. Environmental Protection Agency

iii

Board on Environmental Studies and Toxicology

Commission on Life Sciences

Preface

We are exposed continually to a wide range of chemical substances. Some are known to be toxic at common exposure levels, and others might be toxic. We have special concern about man-made chemicals, which often move readily from place to place—for example, from factory smokestack to air to rain to groundwater to household water supply—and can enter our food supply, air, and soil. Data on the movement and present location of specific chemicals are remarkably sparse, but even extensive monitoring of the concentrations of chemicals in various exposure media often can fail to detect or define human health risks. There are too many chemicals, too many sources, and too many routes of exposure to rely solely on environmental monitoring. Additional problems arise when a chemical is newly recognized as important but was not included in past programs to monitor human exposure, and still other problems arise when the relationships between exposure levels and concentrations in the human body are unknown.

Those concerns make it important to determine the concentrations of specific chemicals in human tissues. The National Human Monitoring Program (NHMP) was established in 1967 by the U.S. Public Health Service and since 1970 has been housed in the Environmental Protection Agency. In response to a request from EPA, this report of the National Research Council's Committee on National Monitoring of Human Tissues evaluates the current program; identifies important scientific, technical, and programmatic issues; and makes recommendations regarding the design of the program and use of its products. The program had not been reviewed in this way at any earlier time, and we believe that it had become a textbook example of a program that was well intentioned, was focused on a critical issue, was managed by staff competent in their disciplines, but was in need of a hard look by external peer review. The program was not large in the overall scheme of things, and peer review would not have been difficult; but by the time our committee was asked to undertake this review, the accumulated problems in the program had reached a stage of crisis, even to the point of doubt about whether the NHMP should be continued. One is moved to wonder how many other small, critically important scientific programs might profit from peer review, and not just at EPA.

The members of our committee were expert, in various combinations, in biostatistics, toxicology, exposure assessment and epidemiology, chemistry, pharmacokinetics, risk assessment, public policy, survey statistics, data base management and tissue archiving, and biologic markers. Before writing this report, the committee convened a workshop in January 1989 to obtain the opinions of program officials and experts in fields relevant to tissue monitoring, environmental monitoring, and risk assessment. These officials and experts helped our committee identify the potential goals and uses of a national program and study in detail the operations and technical methods of the NHMP. Users and potential users of tissue monitoring data also made important contributions to the workshop.

During our work, the committee was repeatedly surprised by the gaps between the needs for data from human tissue monitoring and the limited scope of current activities to fill those needs. Other countries, most notably Germany, have far more extensive human tissue monitoring activities, and the data are used widely for many purposes. In the United States, the right kind of program could generate data of great value to numerous and diverse users. Thus, our main conclusion is that a substantially new program, designed with an appreciation of the strengths and weaknesses of the old, should be implemented forthwith.

The typical committee process often is criticized, but in this case, things worked very well indeed. The members of our committee, all strong and independent professionals, rolled up their sleeves and settled down to work together at our very first meeting. The National Research Council staff were fully effective members of the team. We could not have produced this report without the support of Lee Paulson and Carolyn Fulco; our efforts and deliberations were greatly aided by Jim Reisa, director of the Board on Environmental Studies and Toxicology; and others, especially Karen Hulebak and Shelley Nurse, helped in critical ways. We also profited from the continuing interest and cooperation of present NHMP staff at EPA, its contractors, and many potential users of the data.

John C. Bailar, III
Chairman

Contents

Transition, 162
Summary and Recommendations, 163

Executive Summary

Many toxic substances in the environment, including man-made pollutants, can pose hazards to human health. Monitoring human exposures to these substances can be difficult. Many substances move readily from one environmental medium to another, and reliable monitoring data are sparse for most routes of human exposure. Furthermore, as discussed in the National Research Council's 1991 report, *Human Exposure Assessment for Airborne Pollutants*, monitoring the environment by identifying and measuring concentrations of chemicals in environmental media (e.g., air, water, and soil) is not by itself an adequate basis for assessing human exposures.

Improved understanding of potential health risks has led to a search for indicators of biologic responses to exposure that reveal the progression of events within the body that *lead* to disease *before* disease occurs. Determining the concentrations of specific chemicals in human tissues—such as blood and adipose tissue—can serve in effect to integrate many kinds of human exposures across media and time. A well-designed national program to monitor toxic chemicals in human tissues is a necessary component of an anticipatory strategy aimed at early identification of and response to health and environmental problems concerning xenobiotic toxicants in the environment.

Used as one component of an effort to manage environmental quality and protect public health, tissue monitoring has valuable attributes:

- Tissue samples can reflect exposures accumulated over time.
- Tissue samples can reflect exposures by all environmental pathways and routes of entry into the body, including some that are difficult or impossible to assess by environmental measurement.
- Pollutants in tissue samples have undergone the modifying effects of physiology, metabolism, and biologic availability, and thus can be used to help describe important exposure-response relationships.

1

• Some agents are more concentrated and are, therefore, more readily detectable, in tissue samples than in the environment.

• Tissue samples can offer the opportunity to correlate the tissue concentration of toxicants with other tissue-based biologic markers or indicators of effect that might be predictive of injury or disease within a given person.

Taken together, these characteristics make tissue monitoring a potentially important adjunct to conventional environmental monitoring and one that is uniquely valuable in indicating exposures and doses that may lead to harmful effects.

THE NATIONAL HUMAN MONITORING PROGRAM

The National Human Monitoring Program (NHMP) was established in 1967 within the U.S. Public Health Service to study changes in pesticide residues in the U.S. population. The primary activities of the NHMP are the National Human Adipose Tissue Survey (NHATS) and special studies that support other programs requiring data relevant to chemical exposures. The NHMP, including NHATS, was transferred to the Environmental Protection Agency (EPA) in 1970. Since 1981, the NHMP has been the responsibility of EPA's Office of Toxic Substances (OTS).

The NHATS was redesigned by OTS specifically to identify chemicals to which a representative sample of the U.S. population was being exposed, to establish baselines and trend data on chemicals for toxicologic testing, to identify populations at risk and to set priorities for risk reduction, and to help assess the effects of regulation. To accomplish these goals, the NHATS measures residues of chemicals in human adipose tissue. Since 1967, the NHATS has collected approximately 12,000 samples of adipose tissue—85-90% from autopsied cadavers and the remainder from surgical patients. Tissues are obtained through a national network of pathologists and medical examiners from 47 urban or metropolitan statistical areas (MSAs); no rural areas or small towns outside MSAs are included. In recent years, the number of specimens collected has dropped from an annual quota of 1,370 in the early 1970s to 500-800 samples in the 1980s.

The NHATS has successfully documented widespread and significant prevalence of pesticide exposures in the general population. It also has shown that reduced use of polychlorinated biphenyls (PCBs), DDT, and dieldrin resulted in lower tissue concentrations of these compounds. A trend analysis for 1970-1981 showed a dramatic decline in PCB concentrations after the regulation of PCBs in 1976.

The NHATS program has important and worthwhile goals. When it began some 20 years ago, the NHATS represented the state of the art in pesticide analysis in human tissues. However, while the objectives of the program have grown, program design and support have not kept pace. The NHATS is now out of date and only partially fulfills its objectives. Design and management problems have been compounded by insufficient financial support for the changing and expanding objectives, and the overall quality of the NHATS has deteriorated.

In 1987, EPA proposed to introduce the National Blood Network (NBN) to monitor residues of industrial chemicals in the blood of volunteer donors from three U.S. blood-collection agencies. The NBN was intended to establish baselines and time trends for the nation and for various population groups. It was conceived as a way to complement the NHATS data by permitting less invasive collection of a tissue that reflected more recent exposures compared with the long-past exposures shown by adipose tissue collection. It also would have complemented the NHATS by focusing on volatile organic chemicals (e.g., benzene and trichloroethylene) and elements (e.g., lead, cadmium, and arsenic), as well as semivolatile organics. However, the NBN has not been implemented.

The NHMP has carried out several special studies, alone and with the collaboration of other agencies. Such studies include an archive stability study; a study with the World Health Organization that measured lead and cadmium in blood; a clinical study of PCBs in transformer workers; development of a national body-burden database; and, with the Veterans' Administration, a dioxin-furan study of Vietnam veterans exposed to Agent Orange.

Because of budgetary limitations, the NHATS is the only operating component of the NHMP, and in late 1987, EPA announced its intention to eliminate the NHMP. Congress instructed EPA to retain the program, pending a review of its usefulness, and funds were made available through fiscal year 1990.

CHARGE TO THE COMMITTEE

In response to the congressional request, EPA asked the National Research Council (NRC) to review and evaluate the effectiveness and potential applications of the NHMP. The NRC's Commission on Life Sciences formed the Committee on National Monitoring of Human Tissues within the Board on Environmental Studies and Toxicology. The committee's tasks were as follows:

• Organize a public workshop to identify relevant data and advise EPA on the scientific, technical, and database management issues relevant to the current conduct and future use and development of the monitoring program.

• Review the past applications and future uses of these databases, considering them in the context of the federal government's overall strategy for monitoring the general population for exposures to various toxicants.

• Evaluate these programs in terms of their design, current and potential applicability, and relative priority, to assist the federal government in determining whether or how to continue such networks.

• Evaluate and provide recommendations for current and future resources for the NHMP and assess the validity of various potential uses for which such monitoring data have been proposed, whether from the NHMP or other sources.

• Assess the utility and cost effectiveness of the program and identify major scientific, technical, and ethical issues that should receive priority attention.

• Address the larger issue of developing consensus on biologic markers of environmental hazards in the context of overall federal efforts to monitor markers of human exposure to environmental toxicants.

The first of those tasks was completed January 24-25, 1989, with a workshop held at the National Academy of Sciences in Washington, D.C. Committee members heard from NHATS users, potential users, and persons involved in the operation of specimen and tissue banks. The current report addresses the other five tasks.

RELATIONSHIP OF TISSUE MONITORING
TO EPA PROGRAM PRIORITIES

The goals of the NHMP correspond broadly to EPA's mission in recognition, evaluation, and control of environmental chemical hazards to human health. However, no detailed plan tying the NHMP to specific EPA regulatory programs or research objectives appears to have been developed. Some NHMP results have been used and cited by several EPA programs; e.g., the 1988 *Environmental Progress and Challenges: EPA's Update* cites NHMP results in monitoring markers or indicators of exposures to PCBs, pesticides, and other agents of concern, such as dioxins, in human biologic samples.

Recent EPA planning documents indicate that the NHMP or a successor program could be integrated readily into new research and program objectives. Measurement of exposures in a manner that leads to realistic projections of dose is explicitly recognized in the EPA planning document *Protecting the Environment: A Research Strategy for the 1990s."* In this document, EPA described new initiatives aimed at broadening the research base for agency planning and developing new programs that transcend the previous "end-of-the-pipe" approach to environmental management. The emphasis on human health risk as a focus in regulatory programs specifically indicates the ongoing need and value of a national human monitoring program. Of the top five priorities for new research programs identified by EPA, the third is "development of a national data base on the extent of human exposure to pollution in the U.S." Many of the individual subjects for research identified in the report are related directly to aspects of the present NHMP or to proposed enhancements.

GOALS AND POTENTIAL USES OF A NATIONAL PROGRAM TO MONITOR HUMAN TISSUES

The committee determined that an ideal national human monitoring program should do the following:

• Measure concentrations of known chemical contaminants in human tissues and help identify new or previously unrecognized hazards related to chemical substances found in the environment, especially those resulting from human activities.

• Establish trends in body burdens of toxicants that result from changes in manufacture, use, and disposal patterns, and thus monitor the results of programs intended to control specific chemical hazards.

• Provide biologic samples and data to aid in the evaluation of relationships between environmental exposure and toxic effects for purposes of risk assessment.

• Identify population groups (by age, geographic location, etc.) that might be at risk because of high body burdens.

• Provide data for comparison with results of complementary environmental monitoring programs.

• Provide human tissues essential for research on related matters, such as determination of body burdens; distribution of chemicals among body compartments; identification of biologic markers; and procurement, storage and analysis of human tissues.

• Allow assessment of past exposure to newly identified toxicants.

CONCLUSIONS AND RECOMMENDATIONS

The committee strongly supports a U.S. population tissue monitoring program.

Given the central role of chemicals in modern society, people will be exposed to chemicals. It is prudent that the general population be monitored to aid in assessing magnitudes of exposure and to determine the need for and effectiveness of regulations and other measures to limit risk.

After a thorough review, the committee found that the current NHMP program is fundamentally flawed in concept and execution and should be replaced completely.

The original goals of NHMP (and the NHATS as redesigned) are still valid.

The committee further recommends that a new program of human tissue monitoring be developed with dispatch, that the NHMP be continued only until a successor program is established, and that the change be completed as soon as is consistent with an orderly transition.

Certain aspects of the NHATS should be preserved and evaluated for continued support: the network for collection of adipose tissue specimens (though it will need modification), the tissue archive, and the record of past analyses.

Toxicologic Issues

Although tissue-monitoring data alone can signal the need to conduct studies on specific environmental chemicals, tissue monitoring to indicate past exposures to chemicals in the environment is best viewed as one component of a comprehensive environmental monitoring program. Information on chemical toxicity typically is inadequate; even if it were adequate, planned and unplanned releases of toxic substances and the resulting human exposures still would occur.

The quantities of chemicals present in various media can be used to determine whether the media are potential routes of exposure; however, such measurements are not necessarily reflected in tissue concentrations. Tissue chemical measurements must be supplemented with knowledge of contaminant sources, environmental pathways, environmental concentrations, time patterns and locations of exposure, routes of entry into the body, material toxicity, and latency.

Relevance of Human Tissue Monitoring to Risk Assessment

Risk assessment includes identification of a potential health risk due to exposure to a hazardous agent, determination of sources and magnitudes of exposure to the agent, estimation of the relationship between the potential risk or severity of disease and the dose of the agent, and the integration of this information into estimates of potential risk associated with various exposure conditions. Most risk assessment today is based on estimates of external exposure, which can be used to calculate dose. However, exposure data generally are imprecise and contribute considerably to uncertainty in the risk estimate. Bioavailability—the completeness of absorption of a substance from environmental media—differs according to many factors, including route of exposure and the chemical and physical properties of the substance. Interpretation of tissue concentrations must depend on inherent potency and a broad range of other factors, including exposure to other chemicals. can provide quantitative data on internal dose or biologically effective dose of a chemical or the resulting biologic effects and can theoretically provide greater precision in the risk-assessment process than can the use of crude exposure data.

Choice of Tissues to Monitor

It is not feasible to study a broad range of tissues in a general population sample. Instead, attempts must be made to identify tissues that most nearly account for the body burden of most of the chemicals of concern. When the NHATS was designed, pesticides were of greatest concern, especially halogenated hydrocarbons (particularly halogenated aromatic hydrocarbons), which tend to accumulate in adipose tissues. New trends of environmental exposures, advances in analytic chemistry, increased sensitivity of equipment, and the discontinuation of the use of most halogenated aromatic pesticides and many halogenated aromatic industrial chemicals make the study of other tissues important as well as feasible. The committee considered the collection of blood as well as other tissues and specimens, including lean tissue, hair, urine, and some other biologic fluids.

The present evidence leads the committee to conclude that the basis of a human tissue monitoring program should be broad, random collection of blood samples, supplemented by the continued collection of adipose tissue.

This recommendation includes probability sampling; the needs cannot be satisfied by existing EPA plans regarding the proposed NBN.

Blood collection should be supplemented by the continued collection of adipose tissue, in part to maintain historical continuity while new long-term series of blood measures are established and in part because some important residues are most concentrated in fat.

Measurements of nonrandom samples of adipose tissue will continue to be important for several years, although they might be replaced later with studies of the lipid fraction of blood.

While blood and adipose tissue are being collected, the program should undertake research on how the chemical measure of a xenobiotic toxicant in one tissue is correlated with that in another, so that the effects of nonrandomness in the adipose samples will be better understood, and the continued contribution of the adipose samples (including stored samples) can be evaluated properly.

Regardless of the tissues collected, samples should be accompanied by standardized information on demographics, illness (especially terminal illness), and known occupational or other major exposures to chemicals.

High priority should be given to the collection of matched adipose and blood specimens for future parallel analyses. Matched specimens of fat from different anatomic regions also might be useful.

Sampling Methods

One of the primary deficiencies in the NHATS is that donors of adipose tissues collected are not a representative sample of the U.S. population (see Chapter 4). Most population surveys compromise somewhat on ideal standards; however, departures from probability sampling in the NHATS exceed what most statisticians would consider acceptable. The main deficiencies are these:

• Although the target population is the living U.S. population, the subjects on whom measurements are taken are an uncontrolled mix of recently deceased persons and surgical patients.

• The sample size has been driven by the budget, rather than by needs to satisfy important goals of the program.

• Some important segments of the population are omitted from the sample. The exclusion of the rural population is the most serious omission.

• Although probability sampling was used in the selection of the metropolitan areas that are the first stage of sampling, problems of cooperation forced substitutions for 20% of the areas. Consequently, the extent to which the sample of areas now represents all metropolitan areas is uncertain.

• No sampling method has been designated for choosing the persons from whom specimens are taken. Each medical examiner or pathologist is given a quota by age, sex, and race, but the quotas are poorly adhered to; even if the quotas were met, the quota procedure would be inherently biased.

• There are no specific instructions for pathologists on the body part to be used for specimens. It is implicitly assumed that contaminant concentrations are the same in all adipose tissue in the body.

• Recent uses of composite measurements have made it impossible to provide prevalence estimates and seriously weakened the estimates of mean contamination concentrations for the sex, race, and age subdomains.

• Sampling errors have not been calculated, so users are not informed about the precision of the data.

• There is no plan for regular release of findings to the public.

The basic structure of the NHATS is such that, even with major improvements, its ability to reflect the accumulations of toxic substances for the U.S. population would be seriously limited.

> *If the NHATS were replaced, as the committee recommends, with a blood monitoring program as the primary method of measuring toxic substances in human tissue, the sampling plan should be patterned after the one used in National Health and Nutrition Examination Survey, which includes a close approximation to a probability sample of the U.S. population.*

Such a system also would permit interview with the sample persons to obtain data on covariates.

> *Blood specimens should be collected in strict accord with the method of probability sampling at all stages. The methods used should*

be efficient for giving virtually all persons in the United States a known probability of selection.

During the time in which the NHATS is continued, selection methods should be revised to reduce subjective elements in the choices of counties, hospitals, and specimens.

Collection, Short-Term Storage, and Archiving of Tissues

Many authorities in environmental monitoring believe that a *prospectively* designed tissue bank should accompany environmental monitoring programs, although retrospective analysis is still an important function for an archive. Banked specimens and separately obtained specimens can reveal trends in environmental chemical exposure and identify agents responsible for pathophysiologic changes in humans, plants, and animals.

The goals of a human-tissue bank operated in parallel with an environmental monitoring program should be clear and supported—with a long-term commitment—by the organization sponsoring the program. A specimen bank organized in conjunction with environmental monitoring will have many purposes and must support those purposes. The utility of a tissue bank depends on its correct operation; its resources; and the proper selection, collection, handling, and storage of specimens.

Given the state of the NHATS current archive of tissue, the committee believes that the existing frozen samples of adipose tissue likely have little or no value to a successor program or to other parties.

However, the committee recommends that a successor program give the use of the current archive early consideration, specifically asking whether the archive should be saved indefinitely or discarded, and if it is to be saved, how it should be preserved and used.

The committee advises the archiving of newly collected specimens according to state-of-the-art protocols.

Samples should be collected and stored in a manner that preserves the possibility of basing measurements on individual samples, and a substantial part of the new program should be based on individual analyses. When it can be shown explicitly that values based on individual samples are not needed, some degree of compositing might be appropriate.

Chemical Assay of Specimens

"Monitoring" implies routine measurement that is inherently closed-ended and based on established methods and practices. Requirements and approaches for monitoring programs vary, but usually proceed from a list of analytes (target chemicals) and assay methods that have been validated for the sample type and concentration range of interest. A successful monitoring program maintains results over time for comparison and must, therefore, be technically adequate at the outset (Chapter 6). Comparability is most easily achieved if assay methods are constant. The NHMP includes aspects that are not compatible with monitoring, such as recognition of new agents of concern and detection of chemicals not previously included in monitoring methods. These are appropriate to a population-based biologic surveillance program, but they require an approach to chemical analysis different from that for monitoring.

The committee considered the following aspects of a monitoring program in its deliberations:

• Present knowledge does not permit designation of all substances that might be detectable in tissues or that would be important if detected.

• Present analytic technology is inadequate for surveillance of some chemicals, because of limitations in sensitivity, applicability, and cost.

• The importance of individual target chemicals will increase or decrease over time, but slowly. "Emergency" exposure assessments that might be required when a serious health hazard is discovered would be addressed best by focused special studies.

• Monitoring efforts will provide opportunities for exploration of tissue composition beyond a list of target chemicals. Those opportunities and other efforts parallel to the monitoring tasks should support continuing development of the monitoring program itself.

Overall costs associated with a monitoring program can be reduced in several ways. Some reduce development costs and lead time, some reduce the number of assays needed for monitoring, and some reduce the cost per assay. Reductions in development costs can be achieved by targeting analytes with similar chemical properties to minimize the number or complexity of analytic protocols or by combining method development with actual sample analysis.

Both approaches have some merit, although they also have disadvantages.

> *Once the relative importance of various possible uses of the data has been established, rationales for selecting target chemicals should be incorporated into a systemic weighting scheme and applied as comprehensively as feasible.*

Criteria for determining the relative importance of a candidate target chemical should be separated from issues of analytic feasibility until late in the planning effort. Identification of one or more analytes that might require a new assay protocol would be important in planning future method development.

Design of an adaptable monitoring program with mechanisms for selection of new analytes, development and validation of collection, storage, and assay will permit a monitoring program to remain responsive to current needs and to take advantage of progress in analytic technology in an organized fashion. Many NHATS program decisions appear to have been arbitrary.

> *The NHATS would benefit from regular strategic planning by EPA, frequent agency consultation with program contractors, scientific peer review, and solicitation of advice from interested federal agencies.*

These efforts would help maintain a clear decision-making process that is well documented and understood.

> *Evaluation of data should be a continuing part of program reporting.*

One feature common to all NHATS project reports produced in the 1980s is the nearly complete limitation of analysis of final results to analytic validity. Interpretation of findings in relation to larger program goals (such as time trends, efficacy of interventions, relative importance of different environmental contaminants, and regional or demographic differences in exposures) is an important part of understanding and meeting additional data needs. The committee believes that only NHMP has broad responsibility for making certain that the program is productive in relation to the larger goals.

Analytic methods research is conducted within EPA, at other government programs, and in academe.

> *The NHATS must be able to articulate and, within EPA, influence research priorities for development of new analytic applications of emerging technology and to benefit from new developments.*

The committee doubts that leadership in analysis can be delegated effectively to contracting organizations, and it believes that EPA must maintain substantially more activity and expertise in this regard.

Since 1981, the analytic effort has been modified from year to year and developmental activities have supplanted monitoring to some degree. Including current efforts, data are available from only 2 collection years.

Priority should be given to setting and maintaining a regular schedule for analysis of results of each assay type.

Program Design and Management Issues

Administrative and Agency Issues

Organizational and administrative location of a human-tissue monitoring program in EPA is critical. The selection of an agency to lead national surveillance of chemical exposures is not simple; the multiagency history of the NHMP and the current ambivalence regarding the future of the NHMP are clear indications that the match between program goals, potential benefits, and EPA mandates is not perfect. Although a successful monitoring program must be relevant to regulatory needs, it could and should serve a wide range of client programs, without being dominated by any one.

Unlike many other EPA programs, the NHMP has a purpose and rationale that transcends any individual EPA regulatory objective. The multiple aspects of a national human-tissue monitoring program has meant that, to some extent, the NHMP is not an integrated part of any specific EPA regulatory program and, therefore, is not at the top of any major agenda.

The committee has specific concerns about the untoward effects of placing a monitoring program in any subunit with direct, major regulatory responsibilities.

The committee firmly recommends that monitoring be kept strictly independent of regulation itself.

The committee concludes that considerations of input to policy, impact, visibility, and independence argue for a location at the highest feasible organizational level.

After considering the most likely government units to house a human-tissue monitoring program, and after hearing testimony at the

January 1989 workshop, the committee recommends that a national program to monitor human tissue remain within EPA. However, the location within EPA should be reconsidered, and the committee recommends that it be moved to an organization unit with EPA-wide responsibilities.

A location that is geographically close to other programs and laboratories active in relevant technical disciplines would facilitate important exchanges about methods, as well as followup of findings.

Funding

The critical resources in a program of monitoring human tissues include funding and expertise in appropriate scientific fields. Sufficient funding, with assurance that it will remain adequate over the next few years, is essential. However, the final budget for a program should be determined after the program specifications have been formulated. The major factors involved in determining funding are the annual sample size, the set of chemical assays to be performed, the type of tissue to be collected, the size of the staff needed to monitor the program and analyze results, and requirements for research and development. Funding is flexible to a certain point. However, there is a minimum level of funding below which the program would not be worthwhile, and obviously, the greater the funding, the more detailed the analyses that can be conducted. EPA needs to make a commitment to request at least the minimal funding through the indefinite future.

The committee firmly concludes that support should either be increased enough to support a useful program, or the program should be eliminated.

A clear decision to end the program would discourage present and potential users from expecting data that cannot be produced; would be a clear statement that EPA does not accord human-tissue monitoring a high priority; and would transfer institutional responsibility out of EPA, perhaps to another federal agency.

Serious drawbacks to a decision to eliminate the monitoring program include the likelihood that no comprehensive, coordinated program would be developed elsewhere, loss of skilled staff and institutional memory, and perhaps destruction of specimens that have been banked and saved. The committee urges that EPA consider termination of human-tissue monitoring only under the most compelling circumstances, and even then only after exploration of ways to ensure orderly transfer to another appropriate agency in less-straitened circumstances.

The committee considered possible levels of funding. Funding of $3 million per year appears to be barely adequate to sustain the minimal activity needed to keep a program in long-term existence. The committee does not recommend support at this funding level.

The next support level considered by the committee was $5 million per year, exclusive of staff salaries and overhead. Although the committee did not undertake detailed cost analyses, it believes that EPA's history as well as the operation of other tissue-monitoring programs suggest that $5 million per year could support a substantial flow of high-quality, policy-relevant information. This level still is not munificent support, but it might be sufficient to serve EPA's policy needs and bring some critical distinction to the program. Furthermore, it could be used to develop a solid base of competence, experience, and usefulness to support possible expansion in the future.

Greater financial support—even up to the $25-50 million per year suggested by heads of other agencies—could be put to good use, given appropriate planning and the organizational setting and mission described in this report. However, such allocations do not appear feasible now, so their implications were not explored.

The committee recommends that a full-time program manager and other staff be allocated funds adequate for the planning and full design of a new program concurrent with preparations for the absorption of the NHATS.

At least $3 million per year will be needed to continue current activities and planning. This phase might take 12 months or more.

The committee further recommends that, after the transition period, additional full-time staff be assigned and support be increased to at least $4 million per year for at least 2 years of consolidation.

Science Review

The committee recommends strong and continuing oversight of the new human-tissue monitoring program.

Expert advice can be obtained in several ways, but the numerous criteria required to establish and maintain a human-tissue monitoring program point strongly to a standing outside scientific advisory body *to advise program and EPA management.*

The body should be outside EPA; nearly all members should be knowledgeable about one or more scientific and technical disciplines important to the program. Advice provided by this group would include oversight of program content, program management, program planning, resource needs, technical operations, timeliness, and appropriate dissemination of results. The advisory body should have no other major responsibilities related to the program. The committee envisions quarterly meetings that would taper rapidly to annual meetings.

Other Administrative Issues

Details of program structure and organization depend heavily on a host of management decisions that the committee cannot foresee. Some of the issues that must be considered, however, include in-house scientific and managerial competence; professional staff members fully dedicated to human-tissue monitoring; and program design.

Some professional staff members should be fully dedicated to the program, without competing duties.

The committee specifically recommends that the program be designed in a modular fashion as much as possible to permit critical core activities to be maintained even if other activities must be curtailed, suspended, or ended at some future time.

Analysis and Reporting of Data

A structured approach to basic and more-exploratory data analysis is needed. The plan for each year's monitoring effort should include a data-analysis plan, with analytic subprojects. Well-defined analyses should result in early reporting of findings; analyses that require method development or input that are not in the chemical analysis set might be reported less rapidly or less often. At a minimum, however, descriptive summary analyses of the data should appear with the same frequency as the chemical analyses.

Data from human-tissue monitoring cannot, in general, be fully interpreted without other information regarding tissue concentrations, patterns of exposures, and the metabolism and toxicology of individual chemicals. Reporting of those should be systematized, but this will require close collaboration of statistical analysts with persons providing other information. Comparison data, especially reports of tissue concentrations of chemical agents, should be sought continuously. Strong interagency and intra-agency collaborations in planning and exchange of data will be required to address measurement objections, facilitate the analysis of monitoring-program results, and result in a program that is most useful to cooperating agencies.

The program must produce timely reports regularly.

The committee recommends that at a minimum, an annual report of basic analyses should be produced within a year of completion of the sample collection.

Reports almost certainly will undergo internal review. Extra-agency review of draft reports is desirable, especially peer review by persons on the scientific advisory panel and perhaps other persons with specific expertise.

A human-tissue monitoring program should be designed as a multiple-user service activity. That creates substantial obligations for assisting users to understand what the program does and does not provide, for timely analysis and publication of results, for specific and helpful guidance in access to archived specimens, and for active marketing of products.

A well-defined process for producing a range of outputs is an important part of the planning effort. A schedule should be widely and continually publicized and should be relaxed only under the most compelling circumstances. Scientific staff, who bear most of the responsibility for meeting a schedule, should recognize that timely, high-quality reports on important matters are a sine qua non.

A specific person or persons must be responsible for outreach efforts, which are warranted by the multiuse nature of the program, the wide-ranging interest in the resulting data, and the clear indications that more passive approaches to publicizing program reports have failed to reach some critical target groups.

**Cooperation and Information Transfer
with Other Organizations**

> *A tissue-monitoring and archival program must cooperate and communicate with other branches of EPA, other government agencies, academic and private sectors, and foreign environmental programs. Not only are such cooperation and information exchange important in the operation of human-tissue monitoring, but continuing information exchange will be critical to the efficient operation of the new program.*

The committee thinks that special value might be found in the joint development of a small set of measurements to be made in similar ways across a broad range of programs or a means of establishing comparability among programs that could lead to a worldwide data base for environmental toxicants that persist over long times or migrate across long distances and across national boundaries.

Monitoring Human Tissues for Toxic Substances

1

Introduction

BACKGROUND

Contamination of the environment with chemical substances, including man-made chemicals and pollutants, can pose a threat to public health (Buffler et al., 1985; Upton et al., 1989). The concern can be justified by several examples: In 1984, the National Research Council (NRC) reported that more than 60,000 chemicals[1] were in commercial use in the United States (NRC, 1984a), and the number was growing by an estimated 1,000 per year. About 3,400 pesticide ingredients (active or relatively inert), such as solvents, are registered for commercial use, but data adequate for a complete health hazard assessment are available on only an estimated 10%, and toxicity information was lacking (as of 1984) on almost 40% (NRC, 1984a). Comprehensive monitoring data on industrial chemicals other than pesticides are even less abundant. More than 700 organic chemicals, including 40 carcinogens, have been identified in the U.S. drinking water supply (Harris et al., 1987). Numerous industrial chemicals (including trace metals, polycyclic aromatic hydrocarbons, and volatile organic chemicals) have been detected in ambient air (Hunt et al., 1986). Food can also be a source of exposure to industrial chemicals (NRC, 1987; Gunderson, 1988).

Toxic substances in ambient and workplace air, water, and soil, and in the food supply move readily from one medium to another, and data on the presence of individual industries' chemicals in specific media are sparse. Reliable

---

[1]The environment includes several kinds of toxicants, many of which are single chemicals. Throughout this report, the committee uses the word "chemical" generically; it does not refer solely to industrial chemicals.

monitoring data that include most routes of human exposures are available on only a few chemicals, and as noted in *Human Exposure Assessment for Airborne Pollutants* (NRC, 1991), monitoring the environment by identifying and measuring concentrations of chemicals in various media rarely can characterize human exposures. However, determination of concentrations of specific chemicals in human tissues (such as blood and adipose tissue) is a major tool for integrating human exposures across media and time.

The Public Health Service established the National Human Monitoring Program (NHMP) in 1967 and transferred it to the Environmental Protection Agency (EPA) when that agency was formed in 1970. The NHMP consists of the National Human Adipose Tissue Survey (NHATS), and various special studies that support other programs that require data on chemical exposures. EPA would also like to introduce the National Blood Network (NBN). Because many chemicals that may be harmful to humans are pesticides or other persistent hydrocarbons, EPA's Office of Pesticides Programs operated the NHATS to monitor pesticide concentrations in adipose tissue, where they are likely to concentrate. In 1981, EPA's Office of Toxic Substances (OTS) was given responsibility for the NHATS program and redesigned NHATS to identify chemicals in human tissues, establish baseline data and trends, and identify population groups with unusually high concentrations of toxic chemicals, thus making NHATS more responsive to the needs of OTS. Details of the program's history are in Chapter 2.

CHARGE TO THE COMMITTEE

The National Human Monitoring Program, as developed by EPA, has been funded at decreasing levels, and the NHATS is now the only operating component of the NHMP. In late 1987, EPA announced its intention to delete the NHMP from its budget, although funds were committed through fiscal 1990. Congress, however, instructed EPA to retain the program until its utility could be reviewed. OTS therefore asked the National Research Council to review and evaluate the effectiveness and potential applications of the data collected in the NHMP. The NRC Board on Environmental Studies and Toxicology formed the Committee on National Monitoring of Human Tissues to evaluate the NHMP, to provide recommendations regarding its design and utility, and to identify scientific and technical issues that should receive priority attention. The committee, in keeping with its charge, held a workshop at the National Academy of Sciences in Washington, D.C., on January 24-25, 1989. The workshop was organized to enable committee members to hear from persons who had used NHATS data or tissues, were considered likely to use NHATS data, and who were involved in the operation of specimen and tissue banks. See Appendixes A, B, and C, respectively, for the workshop agenda, participant list, and summary.

ENVIRONMENTAL AND PUBLIC HEALTH RATIONALE FOR MONITORING CHEMICALS IN HUMAN TISSUES

There is general agreement about the need to reduce the release of toxic substances into the environment to protect public health and the environment itself. A corollary is the need to anticipate toxic hazards before they cause disease. A recent National Academy of Sciences report concluded that controlling toxic substances in the environment warrants high priority and challenges both the legal and public health systems of the nation (NRC, 1988). To anticipate and control environmental toxicity, new ways to identify and assess human health hazards in the environment in a timely fashion are needed.

The EPA Science Advisory Board (SAB) has recommended a research strategy for preventing or reducing environmental risks, including those resulting from chemical pollution (EPA, 1988). The strategy includes a long-range research program aimed at characterizing the sources, transport, and fate of environmental pollutants; assessment of total environmental exposure with personal monitoring, models, and biologic markers of exposure; and assessment of human health effects of pollutants with biologic markers of disease, extrapolation of animal effects to humans, and epidemiology. The SAB underscored the need for EPA to improve its capability to anticipate environmental problems, and it cited several instances in which chemical contamination problems (including Kepone in the James River, polybrominated biphenyls in feed, and tributyltin in harbors) were not discovered until after substantial health or economic costs were incurred. The SAB stated (p. 12):

Clearly, great benefit can be derived from the identification of trends in environmental quality before they begin to cause serious ecological or human health problems. . . . EPA needs to begin monitoring a far broader range of environmental characteristics and contaminants than it has in the past. Although we understand a lot about the handful of chemicals that already are known to cause environmental problems, we know relatively little about the thousands of chemicals used in modern society, and that possibly could cause adverse affects on human health and ecosystems over the long term. Thus EPA should expand its use of monitoring activities that can foretell health and ecological risks. Past analysis of the muscles and adipose tissue have provided invaluable information on a wide range of contaminants actually accumulating in living creatures. Those kinds of studies should be increased in the future.

A related recommendation of the SAB was that EPA "expand its efforts to understand how and to what extent humans are exposed to pollutants in the real world" (p. 14). Heightened recognition of potential health risks, which has resulted from ever more sensitive chemical measurement coupled with toxicologic characterization of delayed effects of chronic exposures, has led to a search for indicators of biologic responses to exposure that reveal, before disease occurs, the progression of events that *lead* to disease. Tissue monitoring constitutes a natural link between conventional environmental surveillance and emerging methods of assessing biologic injury that results from exposure to toxic chemicals. For example, detection and measurement of a pollutant chemical in tissue samples would indicate that exposure to it has taken place. A well-designed national program to monitor toxic chemicals in human tissues is directly relevant to identified research needs and is a necessary component of an anticipatory strategy aimed at early identification of and response to health and environmental problems.

Tissue monitoring has valuable attributes if used as one component of an effort to manage environmental quality and to protect public health:

• Tissue samples reflect exposures accumulated over time.
• Tissue samples reflect exposures by all routes, including some that are difficult or impossible to assess by environmental measurement (such as hand-to-mouth ingestion in young children).
• Pollutants in tissue samples have undergone the modifying effects of physiology (rates of uptake, distribution, bioconversion, elimination, and storage) and biologic availability.
• Some agents are more concentrated, and so more readily detectable, in tissue samples than in the environment.
• Tissue samples offer the opportunity to correlate, within a given person, the tissue concentration of toxicants with other tissue-based biologic markers or indicators of effect that might be predictive of injury or disease.

All those characteristics, taken together, make tissue monitoring as an assessment tool an important adjunct to environmental monitoring that is uniquely valuable in indicating both exposures and doses that lead to potentially harmful effects.

RELATIONSHIP OF TISSUE MONITORING
TO EPA PROGRAM PRIORITIES

During the early years of the NHATS, few planning documents explicitly stated EPA's view of the relationship between national human tissue monitoring and the agency's responsibilities under various legislative mandates. In its 1976 report to Congress (*Environmental Research Outlook for FY 1976 Through 1980*), the Office of Research and Development did not explicitly refer to human exposure and assessment or to monitoring of biologic samples, although the need for a national plan for monitoring pesticides was identified under the Federal Insecticide, Fungicide, and Rodenticide Act (FIFRA). EPA's mission in recognition, evaluation, and control of environmental chemical hazards to human health corresponds broadly to the goals of NHMP in surveillance of trends in human exposures to chemical hazards in the United States, but no detailed plan for tying the NHMP to specific EPA regulatory programs or research objectives appears to have been developed. Some specific NHMP results have been widely used and cited by several EPA programs. *Environmental Progress and Challenges: EPA's Update* (EPA, 1988) cites NHMP results in monitoring exposures to polychlorinated biphenyls, pesticides, and agents of more recent concern, such as dioxins, in human biologic samples.

More recent planning documents indicate that the NHMP or a successor program could be readily integrated into new research and program objectives. The role of population-based human tissue monitoring in the overall context of EPA's mission and current goals lies in providing a link between environmental concentrations of pollutants in the various exposure media and biologic consequences that might be predicted or observed. The necessity for measurement of exposures in a manner that leads to realistic projections of dose is explicitly recognized in the EPA planning document *Protecting the Environment: A Research Strategy for the 1990s* (EPA, 1990). In citing emerging environmental management issues including the need for multimedia management of toxic chemicals and the apparent disparity between relative health risk and regulatory priorities, EPA has announced new initiatives aimed at broadening the research base for agency planning and developing new programs and at transcending the previous compartmentalized "end-of-the-pipe" approach to environmental management. The new emphasis on human health risk as a unifying focus in regulatory programs for air quality, drinking water, waste management, toxic chemicals, pesticides, ecologic protection, etc., would specifically indicate the ongoing need and value of a national human monitoring program. Of the top five priorities identified by EPA for new research programs, third in importance is "development of a national data base on the extent of human exposure to pollution in the U.S." (p. 2). Many of the individual subjects for research identified in the report are related directly to aspects of the present NHMP program or to proposed enhancements: development of biologic markers of exposure, development of techniques for the assessment of human exposure, and reliance on indicators of internal concentrations or quantities of pollutants (doses) rather than external quantities (environmental concentrations), for risk assessment purposes.

In summary, the relevance of national human tissue monitoring to EPA's mission has always been identifiable. Recent strategic planning to identify EPA's priorities for the next decade give human tissue monitoring a more central role than it had in the past and would require that some aspects of the NHMP be created, if they did not already exist.

GOALS AND POTENTIAL USES OF A NATIONAL PROGRAM TO MONITOR HUMAN TISSUES

An ideal national human monitoring program should:

• Measure concentrations of known chemical contaminants in human tissues and identify new or previously unrecognized hazards related to chemical substances, especially those attributable to human activities.

• Establish trends in body burdens of toxicants that result from changes in manufacture, use, and disposal patterns, and thus monitor the results of programs intended to control specific chemical hazards.

• Provide biologic samples and data for the evaluation of relationships between environmental exposure and toxic effects for purposes of risk assessment.

• Identify population groups (by age, geographic location, etc.) that might be at risk because of high body burdens.

• Provide data for comparison with results of complementary environmental monitoring programs (e.g., NOAA's Mussel Watch Program and National Institute of Standards and Technology (NIST)'s Environmental Specimen Banking Program).

• Provide human tissues essential for research on related matters, such as determination of body burdens; distribution of chemicals among body compartments; identification of biologic markers; and procurement, storage and analysis of human tissues.

• Allow assessment of past exposure to newly identified toxicants.

The products of such a monitoring program—reports of tissue concentrations of contaminants representing groups of the U.S. population—will stimulate the formulation of questions regarding patterns of exposure and other exposure-related issues. Those questions might not be answerable with existing information, but they can focus regulatory or scientific attention on new problems. Estimates of exposures based on tissue concentrations of contaminants (discussed in Chapter 3) can contribute to quantitative risk assessment of detected agents or of nondetected agents with known analytic detection limits.

> *The committee finds that a program of human tissue monitoring is critically necessary to continuing improvement of understanding of exposure to toxic chemicals and recommends that such a program be given high priority for funds and other resources.*

Most of this report expands on that recommendation.

2

Review of the National Human Adipose Tissue Survey and Selected Program Alternatives

DEVELOPMENT OF THE PROGRAM

The National Human Monitoring Program (NHMP) was established in 1967 to determine and assess changes in detectability and concentrations of pesticide residues in the national general population. The program was initially an activity of the U.S. Public Health Service, but was transferred to the new Environmental Protection Agency (EPA) in 1970. The NHMP now consists of the National Human Adipose Tissue Survey (NHATS); EPA planned to introduce a National Blood Network (NBN), but it has not been implemented.

It is difficult or impossible to remove persistent xenobiotics effectively from the tissues of living animals or people, so precautions must be taken against unnecessary and excessive exposures. Precautions against exposure are, however, greatly complicated by the dependence of modern society on synthetic chemicals; even the cleanest-looking environments now contain toxic contaminants. Therefore, it is prudent to assess and try to reduce their effects. An essential part of the management of toxic chemicals is regular sampling of the population to determine magnitudes of exposure and contamination.

Neither detection nor determination of concentration of a chemical in tissues of individuals or the general population provides a quantitative estimate of risks to human health. For example, people living downstream from a defunct DDT plant in Triana, Alabama, were chronically exposed to DDT in their diets for many years and have several times the national geometric mean for DDT and related chemicals (e.g., DDD and DDE) in their serum. However, a survey of the health status of 499 of them, out of approximately 600, indicated that total DDT body burdens were not associated with specific illness or ill health (Kreiss et al., 1981). But some chemicals can be degraded so rapidly that they cannot be detected in tissues even a few hours after exposure (Lynn et al., 1984). Such chemicals include benzidine (a known human carcinogen), aflatoxin, vinyl chloride, formaldehyde, and many other toxic or carcinogenic chemicals. Therefore, failure to detect a chemical in human tissues does not mean that potentially toxic exposure to it has not occurred.

The NHATS was intended to provide data on the existence and concentrations of chemical substances in human adipose tissue of a representative sample of the U.S. population. Its specific objectives were to identify chemicals to which the population was being exposed, to establish baseline and trend data on detected chemicals, to identify populations at risk and set priorities for risk reduction, and to determine the impact of regulation. The NHATS has always measured residues of chemicals in human adipose tissue. Adipose samples are submitted for chemical analysis through the cooperation of selected pathologists. Those samples are collected from postmortem examinations and from remnant specimens removed during surgery. Age, sex, race, and clinical pathologic diagnosis are requested for each patient sampled. Geographic area of residence is inferred from the location of the contributing hospital.

The NHATS originally targeted polychlorinated biphenyls (PCBs) and several organochlorine pesticides for study. Those compounds are highly lipophilic and stable in biologic systems and therefore accumulate in adipose tissue; in fact, human adipose tissue contains the highest concentrations of some of the most persistent chemicals to which humans have been exposed. The NHATS therefore selected adipose tissue as best for its purpose of estimating past human exposure. Although the emphases of the program and the chemical classes targeted by the NHATS have changed substantially, adipose tissue continues to be the basis for study.

Since 1967, the NHATS has collected approximately 12,000 samples of adipose tissue—some 85-90% from autopsied cadavers and the remainder from surgical patients. Tissues are collected through a national network of pathologists and medical examiners from 47 urban or metropolitan statistical areas (MSAs); no rural areas or small towns outside MSAs are included. In recent years, collection of specimens has dropped from an annual quota of 1,370 to 500-800 samples.

The NHATS specimens collected during fiscal year 1982 were designated for "broad scan analysis" to determine volatile and semivolatile organic compounds (EPA, 1986a). Since 1982, the program has been modified to implement newer methods of detecting chemicals on the original target list and to expand the list of chemicals that can be detected. New methods have come into use to measure PCB homologs, chlorinated benzenes, aromatic hydrocarbons, phthalates, and phosphate esters. Those changes signaled a shift in emphasis from monitoring to exploration and the replacement of some routine assay techniques with newer methods. Initially, individual samples were assayed; however, because of the high cost of assays, EPA later switched to the use of composite samples. Forty-six composite samples were prepared from more than 750 NHATS specimens collected during fiscal year 1982 according to a study design prepared by the EPA Office of Toxic Substances (OTS) design and development contractor, Battelle Columbus Laboratories.

The NHATS data, reported in a series of publications, have documented a widespread and significant prevalence of pesticide residues in the general population (Kutz et al., 1979; Strassman and Kutz, 1981). The program has also identified a high-risk population exposed to the pesticide Mirex (Kutz et al., 1985). NHATS data have shown that reductions in use of PCBs, DDT, and dieldrin have been followed by a decline in measured concentrations of these compounds. A specific high point for the NHATS came in 1982 with the release of a 1970-1981 trend analysis showing a dramatic decline in PCB concentrations after the 1976 PCB regulation. With this as an impetus, EPA moved to develop state-of-the-art analytic protocols.

In fiscal year 1987, EPA proposed a National Blood Network that would use three U.S. blood-collection agencies as a surveillance system to monitor residues of industrial chemicals in the blood of volunteer donors. The purpose was to establish baselines and time trends for the nation as a whole and for various population groups. It was intended to complement the NHATS data by permitting less invasive collection of specimens from some subjects (blood rather than fat) and to reflect more recent, rather than only long-past, exposures. It would also complement the NHATS by focusing on volatile organic chemicals (e.g., benzene and trichloroethylene) and elements (e.g., lead, cadmium, and arsenic), as well as semivolatile organics.

The NHMP has also carried out several special studies, alone or in collaboration with other agencies. They included an archive stability study, participation in a World Health Organization study of lead and cadmium in blood, a clinical study of PCBs in transformer workers, development of a national body-burden database, and (with the Veterans' Administration) a dioxin-furan study of Vietnam veterans exposed to Agent Orange.

The NHMP has suffered in recent years because of budget cuts. The NBN has never been initiated, and analysis of NHATS specimens has lagged. The NHMP is now funded only through fiscal year 1990.

Program Deficiencies

The NHATS program is the only comprehensive human tissue monitoring program in this country. In the face of the data needs generated by the vast use of chemicals in the United States, it is expected to provide a national assessment of chemical exposures on an annual budget of less than a million dollars. In spite of its small present budget, it has collected approximately 12,000 samples of human adipose tissue over some 20 years through a national network of pathologist and medical examiners. EPA and its dedicated staff deserve recognition for that effort.

However, the program has a number of serious problems. A basic statistical problem is that the original design was very weak; it used probability sampling for only one stage of sample selection, and even that use has been diluted over the years. Thus, the ability to generalize results is seriously limited. Another weakness is that the original design of the program did not include rural areas in the samples. The annual quota for sample collection has been 1,370; however, the average has been much lower since 1970, and only 500-800 samples have been collected in part because of a decrease in funding. Another concession to decreased funding has been a move to assay composite, rather than individual, samples. Budget reductions have also caused proposed modification and expansion of the program to be delayed, with indefinite postponement of such key steps in the program as chemical analysis of tissues, data analysis, methods development, and reporting of survey results. And the blood network has yet to be implemented.

The NHATS program is now more than 20 years old. It has not aged or developed gracefully. Collection and analysis of samples have not been nearly as effective or as representative of the population as planned. Storage of samples has been inappropriate, and there is a question of whether samples collected more than a year or two ago can be used for adequate quantitative estimation of donor exposure; such specimens might be only minimally adequate even for qualitative analysis of some exposures. Data obtained from the samples have been the basis for several publications and reports, but they have not been used as effectively as they could have been. Relative to the time and money invested in the program, the publications and reports are few and cursory; and publication of reports describing survey results is usually several years behind data collection. All those factors have combined to damage the prestige and credibility of the program. The NHATS has rarely enjoyed the support that such a program deserves, and, because of its waning reputation, it currently appears to be a program that higher EPA management wishes to avoid rather than support.

In summary, the NHATS program was well intended and has many important and worthwhile goals. At its inception, approximately 20 years ago, the program was at the state of the art of pesticide analysis in human tissues. However, NHATS management has expanded the objectives of the program to encompass more than was originally intended while failing to improve and expand its design and support appropriately. The lack of innovation and leadership over the years has resulted in a program that is out of date and only partially fulfills its objectives. Design and management problems have been compounded by the failure of the program to receive financial and managerial support in proportion to its expanding objectives. As a result, the program has promised more than it can deliver, and both the quality and the quantity of data obtained from the NHATS have failed to meet reasonable expectations. Without the necessary managerial and financial support, the overall quality of the NHATS has deteriorated to the point where questions have been raised as to whether it is still worth while or even salvageable.

A short summary of the major weaknesses of the current NHATS program follows. More detailed discussion of these weaknesses is presented in later chapters.

Toxicologic Issues (see Chapter 3)

A blood collection system has been proposed, but all prior tissue collection has been limited to adipose tissue. Variation of adipose storage of chemicals with body site (e.g., subcutaneous vs. perirenal) has not been reviewed carefully. And analysis of chemicals with lower lipid partition coefficients (e.g., heavy metals) has not been emphasized.

Storage of specimens under suboptimal conditions might prevent the analysis of specific biologic markers; for example, some may be unstable at the storage temperatures now used.

The current program has failed to correlate and evaluate tissue concentrations of chemicals with other variables important to interpretation of the concentrations (e.g., data on exposure and toxicity).

Sampling Strategy (See Chapter 4)

There are serious problems with the NHATS sample design. Some of them are inevitable when adipose tissue is used as the unit of analysis, and others follow from decisions made in planning the study. The most severe problems areas follow:

• The population of interest is presumably the entire live U.S. population, but the measurements are made on a combination of adipose tissue from recently deceased persons and surgical patients (as collection of adipose tissue is necessarily a invasive procedure); tissues from these sources may be substantially affected by serious illness, treatment, and/or cachexia with concentration of fat-soluble substances. There has been no evaluation of whether contaminant concentrations in tissue from these sources are similar to concentrations in the population as a whole.

• Most of the rural population is excluded from the sampling frame; there is no information on urban-rural differences that would permit inferences about the U.S. population.

• Probability sampling was used for only one of four stages of sampling—the selections of metropolitan sample areas. There is no information on potential biases of the sample selection. In the one stage that was selected with probability methods, there has been a deterioration in the quality of the sample.

• There has not been adequate consideration of the precision needed for data analysis or of the sample sizes appropriate for the NHATS.

Insufficient attention has been paid to the implementation of the sample selection procedures. As a result, the sample sizes specified for the NHATS have not been attained in recent years. Furthermore, the contractor's latitude in choosing counties and medical examiners or pathologists and the medical examiners' and pathologists' latitude in choosing tissue specimens are so wide that it is not clear how well the specimens represent the population that was intended by the sampling protocol.

Starting with samples collected in 1982, the NHATS stopped separate analyses of each tissue specimen and introduced analyses of composites; that led to a substantial loss of ability to analyze and interpret NHATS results. There is no direct way to derive prevalence estimates. Statistical modeling is used to estimate mean concentrations by age, sex, and race, but the adequacy of the models has not been tested sufficiently.

Little or no effort has been made to establish confidence levels around the NHATS estimates. As a result, analysts have difficulty in knowing whether changes over time or differences among population groups reflect real differences or are only random fluctuations.

Collection, Storage, and Archiving
of Tissues (see Chapter 5)

Tissue samples now stored have been transferred among several repositories, and some have been thawed (both accidentally and deliberately to permit partial analysis). There is a lack of storage history, including freeze-thaw history, of many earlier specimens.

Inadequate numbers of specimens are being collected, and storage conditions at collection sites are not uniformly controlled. Specimen storage containers are not suitable for some analytes.

Collection sites have no organized quality control system. Training programs for collectors are lacking, site visits are not made to collectors, and exclusion criteria on samples are not enforced. Many tissue specimens appear to be seriously compromised, and there is evidence of sample contamination.

Sample holding times of several years have been accepted, without studies validating the storage stability of target compounds over those times.

Specimens in the archive are stored in an unorganized manner. There is evidence that sample containers have been inappropriate.

Chemical Assay Issues (see Chapter 6)

The list of compounds selected for monitoring has not been regularly reviewed and modified, and it has been expanded without sufficient systematic planning.

Analytical methods introduced in 1981-1985 were implemented without sufficient validation, and analytic goals are not clearly defined or defended.

Routine monitoring functions have been allowed to lapse, because of sporadic analyses, compositing of samples, and lack of continuity of results.

Programmatic Issues (see Chapter 7)

A critical aspect of a successful monitoring program is the reporting of findings in a format and with a timeliness that match the needs of data users. For example, tracking of a rapidly changing environmental contaminant may require regular reporting to a regulatory or enforcement office, with a delay of no more than 6 months after the close of each reporting year. Reporting of monitoring data from the NHATS was first delayed, was then at a reduced frequency, and finally broke down altogether in the late 1980s. Part of the problem was excessive EPA delay in that part of EPA that was reviewing contractor-prepared reports.

Large programs can be contracted out successfully, but the export of functions must be closely matched to both the export of responsibilities and continuing technical oversight. It is the committee's impression that contract work in support of the NHATS has been performed in a technically acceptable manner, but that critical decisions about the general thrust of the contract work have not had adequate attention from either EPA or the contractors.

Examples (cited above) include conditions of collection and storage of samples, choice of substances to be examined, and laboratory methods. In short, the contractors appear to have done an acceptable job, but sometimes the wrong things were measured.

Little evidence can be found of long-range strategic planning by the staff. Program staff were concerned with such matters as maintaining the projected magnitude of specimen collection, but failed to attend to long-range planning that would enhance the value of the collected specimens. For example, there were apparent failures to build strong links with the community of potential users, to examine and deal with the implications of the program's organizational location, to plan for adequate continuing input from a range of technical specialties, to monitor programs in other countries and to develop close communication and even working relationships with other programs in EPA, other federal agencies, and the private and nonprofit sectors.

The present program has suffered, at first indirectly and later directly, from a failure to give substantial and explicit attention to promoting the use of its products (i.e., the adipose tissue archive and data from the analysis of the samples). As a result, the products have been underused or even unused. A symptom is the recurring reference to a few striking figures of the decline in DDT, PCBs, and dieldrin after regulation of use (GAO, 1988). The committee acknowledges both the importance of and the public interest in those examples, but there should be by now a large number of additional examples of the importance of NHATS data.

The NHATS as a program has suffered severely from a lack of full-time, concerted attention from persons who are technically competent in the specific relevant disciplines and who have substantial leadership roles in EPA. The committee is pleased to acknowledge the individual abilities of NHATS staff, but their commitment is far short of full-time because of other, competing responsibilities. Furthermore, the program has not been supported by sufficient leadership or full-time managerial personnel.

EXISTING PROGRAMS AS POSSIBLE ALTERNATIVES TO THE NATIONAL HUMAN ADIPOSE TISSUE SURVEY

The committee made a deliberate effort to consider alternatives to the present or a related successor program. This section briefly describes several programs that have some points of similarity to the National Human Monitoring Program and considers whether the NHMP objectives might be met with data from these programs. The major criteria for evaluating the suitability of each of these programs to assume NHMP tasks are shown in Table 2-1. The

TABLE 2-1 Extent to Which Existing Programs Meet Criteria for a National Human Tissue Monitoring Program

Criteria	NHATS	NHANES	ATSDR	NIST	TEAM	FDA	NOAA
Population-based sample (can describe the U.S. population and major groups)	No	Yes	No	No	No	No	No
Long-term commitment to monitoring (regularly repeated observations with sufficient frequency to detect changes)	Yes	No	No	No	No	Yes	Yes
Focus on toxic xenobiotics (can respond to chemicals in commerce)	No	No	Yes	No	No	Yes	Yes
Relevance to emerging research in toxicokinetics, exposure assessment, and risk assessment	No	Yes	Yes	Yes	Yes	No	Yes

committee, therefore, considered substituting existing programs as alternatives to NHMP if the present program is terminated.

National Center for Health Statistics:
National Health and Nutrition
Examination Surveys (NHANES)

The committee has given special thought to whether the National Health and Nutrition Examination Surveys (NHANES) of the National Center for Health Statistics (NCHS) could be adapted (e.g., by collecting additional blood samples) to serve national needs for human tissue monitoring of contaminants. Because the NHANES already takes blood samples, NHANES analysis of blood might seem to be a plausible alternative to a separate program. Our review of the NHANES indicates that it is not an adequate substitute. The NHANES has some attractive features: a national probability sample, the continuing collection of blood, the availability of other health and personal information on each subject, and a strong, broad staff that is already skilled in many relevant disciplines and techniques. Other strengths include the probability estimates of various features of the U.S. population that might have health significance or be useful in looking at health risk factors, and a priority toxicant reference-range study being conducted as part of NHANES III in collaboration with the Agency for Toxic Substances and Disease Registry. The study is to determine background concentrations of 50 priority toxicants, primarily volatile chemicals and phenols, in 1,000 people with no known excessive exposures. Subjects are being selected by age, sex, and geographic region. Urban-rural status will be considered. (See Appendix D for a more complete discussion of the NHANES.)

A combination of four other features, however, seems fatal to the use of the NHANES for the present purpose:

- The national need is for regular, continuing sampling of human tissues, whereas the NHANES is periodic (recently at intervals of 5-10 years). That is not often enough to identify new trends rapidly.
- The NHANES has in the past been slow in providing results to sponsoring agencies (personal communications and workshop presentations), whereas a properly functioning monitoring program should provide results quite promptly—indeed, substantially more promptly than the present NHMP.
- Blood taken for the purpose of human monitoring must be in substantial volume (probably 200 cm^3 or more), and such a volume could not be added to the 150 cm^3 NHANES already collects.

- The NHANES has, properly, a range of competing goals, and those goals might change from time to time in response to changing perceptions of national needs in ways that would not support tissue monitoring. Thus, the NHANES cannot have a primary commitment to NHMP goals and is not suitable vehicle for human tissue monitoring.

Agency for Toxic Substances and Disease Registry: National Exposure and Disease Registries

The Agency for Toxic Substances and Disease Registry (ATSDR) was created and directed by Congress to implement the health-related provisions of three laws that are designed to protect the public from adverse health effects of hazardous substances: the Comprehensive Environmental Response, Compensation, and Liability Act of 1980, amendments to the Resource Conservation and Recovery Act of 1984, and the Superfund Amendments and Reauthorization Act of 1986. The agency has developed 10 programs to aid in the implementation of its congressional mandates. Three of the programs (in health assessments and health studies, toxicologic profiles, and exposure and disease registries) are to evaluate the adverse human health effects and diminished quality of life resulting from exposure to hazardous substances in the environment. Although ATSDR has been considered as a possible parent agency for a new human tissue monitoring program (Chapter 7), it does not have in-house capability to collect and assay human tissues. Therefore, transferring the NHMP to ATSDR would be equivalent to, rather than an alternative to, planning a new program.

National Institute of Standards and Technology: Environmental Specimen Banking Program

The National Institute of Standards and Technology (NIST) carries out environmental-specimen banking and archiving. The specimen-banking activities include continuing cooperative efforts with the National Oceanic and Atmospheric Administration (NOAA). Approximately 10 years ago, NIST (then known as the National Bureau of Standards) began collaboration with EPA to determine the feasibility of long-term storage of environmental specimens. The EPA Office of Research and Development and Office of Health Effects Research funded a pilot environmental-specimen banking program.

NIST's approach was to gain experience in the various aspects of specimen banking, including collection, storage, and analysis. Although the program was initially intended to focus on four kinds of samples—human, marine, food, and air samples—funds have been available only for the study of human tissues and, to a smaller extent, marine samples (e.g., in cooperation with NOAA). Collection and banking of human liver specimens began in late 1979. More than 550 specimens have been collected for trace elements and organic pesticides, and PCB measurements are available for about 100 of these samples.

NIST clearly is capable of directing the development of human tissue monitoring programs and could contribute substantially to improving the technology of sample collection, storage, and assay. However, routine monitoring programs do not appear compatible with NIST institutional goals. The cost constraints and requirements for maximal numbers of samples to be assayed would, in the opinion of this committee, present an insoluble conflict with other NIST priorities and functions.

Environmental Protection Agency: Total Exposure Assessment Methodology Study

EPA has carried out the Total Exposure Assessment Methodology (TEAM) study to measure the personal exposures of 600 people to particular chemicals that were selected on the basis of their toxicity, carcinogenicity, mutagenicity, production volume, presence in preliminary sampling and pilot studies, and amenability to collection on Tenax. The subjects were selected to represent a total population of 700,000 residents of cities in New Jersey, North Carolina, North Dakota, and California. Each participant carried a personal air sampler throughout a normal 24-hour day and collected 12-hour daytime and 12-hour overnight urine samples. Identical samplers were set up near some participants' homes to measure ambient air. Each participant also collected two drinking-water samples. At the end of the 24 hours, each participant contributed a sample of exhaled breath. The air, water, and breath samples were analyzed for 20 target chemicals.

The products of a rejuvenated national tissue monitoring program would be highly complementary to such studies of exposure dynamics. The EPA personnel and organizational units that have been associated with TEAM studies might well be candidates for involvement in a successor NHMP. However, TEAM studies themselves are not an alternative to national human tissue monitoring. The series of detailed portraits of exposure (principally to organic vapors) represented by TEAM studies are distinct from the long-term tissue monitoring snapshots that would be provided by an updated National

Human Monitoring Program. Some of the key differences are the populations represented, the chemical agents addressed, and the exposure time scale represented by biologic samples.

Food and Drug Administration: Total-Diet Study

The Food and Drug Administration (FDA) monitors the U.S. food supply for pesticide residues, toxic elements (including heavy metals), and industrial contaminants. The objectives of the program have varied over time, but enforcement of the pesticide tolerances established by EPA and determination of the incidence and concentrations of chemical residues in food are of primary concern. Continuing studies are directed toward those objectives. For example, FDA has carried out a large-scale monitoring program for pesticide residues since the early 1960s. The program has two principal approaches: a commodity monitoring program to measure residues in specific domestic and imported foods and to enforce tolerances and other regulatory limits and a total-diet study to measure the intake of pesticides in foods prepared for consumption.

The Total-Diet Study (also called the Market Basket Study) would not serve as a substitute for tissue monitoring, given other important sources of exposures and the incompleteness of present information regarding the biologic fate and diversity of chemicals that might be found in the diet. The total-diet study, like the TEAM study, would complement a national program for tissue monitoring, but it is not a substitute.

National Oceanographic and Atmospheric Administration: National Status and Trends Mussel Watch Program

The National Status and Trends program has specific, well-defined goals to quantify the spatial distribution and long-term temporal trends of contaminants in the marine environment. The goals are accomplished by an annual collection of bivalves, livers of bottom-feeding fish, and surface sediments in coastal and estuarine areas of the United States. NOAA's NS&T Mussel Watch collects specimens from 150 sites around the United States, and the NS&T Benthic Surveillance Program collects specimens from 50 other sites. NIST is responsible for the quality assurance of organic analyses; quality assurance for measurements of trace elements is handled by the National

Research Council of Canada. There is also a specimen banking component of the National Status and Trends program, which is carried out with the cooperation of NIST.

Environmental monitoring programs in general and marine tissue monitoring data specifically do not address the full scope of objectives of a National Human Tissue Monitoring Program.

Foreign Programs

The committee reviewed several foreign monitoring programs. A comprehensive program of environmental monitoring that uses human tissues has been developed in Germany. That program and other foreign programs are described in Appendix E.

Conclusions

The committee concluded that the approach of each of the programs just described plays an important role in the identification and control of hazards to human health. Coordination and cooperation among those programs would benefit researchers, health professionals, and the public; such an effort also would enhance the federal approach to monitoring public and environmental health. However, none of the programs meets all the criteria in Table 2-1. Not all can incorporate new research findings in toxicokinetics, exposure assessment, and risk assessment. Not all can take the place of human tissue monitoring, which should offer over the long-term, unique insights into substances that actually enter human tissues and in what amounts. For example, only some of the other approaches could measure average concentrations of lipid-soluble toxins (such as PCBs and dioxin) or heavy metals (such as lead and mercury) to determine whether any demographic group might have exposures of concern.

SUMMARY AND RECOMMENDATIONS

After a thorough review of the present NHMP program, the committee finds that NHMP is fundamentally flawed in concept and execution, and should be replaced in toto.

In discussions within this committee, in comments from experts during the workshop and in separate discussions, and in the technical materials we reviewed, the need for a U.S. population tissue monitoring program is supported without reservation.

The original goals of NHMP are still valid, especially as the widespread use of chemicals in modern society continues to increase. Every reasonable effort should be made to protect the population from the untoward effects of chemical exposures. A vital part of the protection is an assessment of past and present exposures to pesticides and other potentially toxic chemicals with a new program that takes advantage of the latest developments in analytic chemistry, statistical design, and human health risk assessment and is adequately supported, both financially and managerially. Although the committee is aware that the rapid metabolism of some pesticides and compounds (e.g., dichlorvos, which is difficult to detect in tissues even after high therapeutic doses are given) will prevent exposure determination for all compounds, the program still will provide valuable data on chemical exposures.

In summary, the committee recommends that a new program of human tissue monitoring be developed.

The committee further recommends that such development be completed with dispatch, that NHMP be continued only until there is a successor program, and that the change be completed as soon as is consistent with an orderly transition.

Aspects of NHMP that should be preserved and evaluated for continued support include the network for collection of adipose specimens, the tissue archive, and the record of past analyses. Later chapters offer suggestions as to how a new program could be designed to meet the objectives of providing data suitable for the accurate assessment of exposure of both the general population and selected groups to a variety of chemicals likely to be encountered in the workplace, through environmental exposure, or in the food supply.

3

Toxicologic Issues

INTRODUCTION

Tissue monitoring for chemicals in the environment is best viewed as a component of a comprehensive environmental monitoring program. This chapter focuses on the relation of analytic measurements of chemicals in tissues to broad toxicologic issues. Tissue-monitoring data alone can alert those concerned with public health to the need to conduct studies on specific environmental chemicals. Three examples that illustrate the importance of tissue monitoring are monitoring of blood to determine the extent of lead toxicity in the United States, monitoring of fish and wildlife to determine concentrations of pesticides in tissues, and monitoring of acute tissue damage to identify conditions common to such damage.

Such examples show that tissue monitoring can be described as a component of an environmental-health and public-health program. Tissue monitoring can reveal some of the associations between the entry of a chemical into the environment and an adverse effect on human health.

RELATION BETWEEN ENVIRONMENTAL MONITORING
AND TISSUE MONITORING

Knowing the concentrations of a chemical in exposure media is helpful in determining the potential risk associated with exposure to it, but such information must be supplemented by information on exposure itself. For example, assume that the concentration of a chemical in some medium of interest is 1 ppm. Obviously, the appropriate magnitude of concern will depend on whether consumption of the medium is, say 5 ml/day (about 1 teaspoon/day) or 2

L/day (about 2 qt/day). Estimates of exposure typically include a concentration component and a consumption component, or equivalent.

For many chemicals of interest, exposure is not limited to one medium. Pesticides can be sprayed directly on crops, but runoff water enters the water supply and spray can travel by air to enter the food supply remote from the point of application. Routes of exposure to chemicals with diverse commercial applications (e.g., lead) can include air, water, food, cosmetics, and drugs, both prescription and over-the-counter. Measurement of exposure through one of many media cannot always provide a full picture of human exposure. Tissue monitoring, however, yields a measurement of the amount of a chemical transferred into an organism, whatever the source and medium. To be transferred into tissues, the chemical must be absorbed, i.e., it must pass through various membranes, such as intestinal membranes, the alveolar membrane of the lung, the conjunctival membrane, and the skin.

The completeness of absorption of a substance from environmental media is referred to as bioavailability. Bioavailability differs according to route of exposure, chemical and physical properties of the substance, and physiologic and nutritional status of the organism. Furthermore, many chemicals that are rather benign in the form in which they exist in the environment might be converted to a more toxic form when they are absorbed.

Clearly, tissue monitoring can be of great value for associating adverse health effects with exposure to environmental chemicals, but several problems arise in attempts to understand the association. Most toxicologic concerns are related to chronic diseases that can take decades to develop (e.g., dementias, neurologic disorders, cancer, osteoporosis, and arthritis). The environmental conditions that may have been significant in the development of a disease and the mechanisms of most of those diseases are not completely understood (e.g., those diseases might be produced by infections or injury).

Given the complexity of chronic disease processes, it is sometimes difficult to associate the onset and development of a case of a disease with a specific chemical. The association of some chemicals with a disease process is well accepted (e.g., appearance of specific proteins in familial Alzheimer's disease); where the association of others is not definite (e.g., DNA adducts). Exposure to some chemicals that do not definitely produce disease (or do not produce it through an understood mechanism) might nevertheless lead to the appearance of some effect. Such an effect is commonly referred to as a biologic marker.

RELEVANCE OF HUMAN TISSUE MONITORING
TO RISK ASSESSMENT

The process of predicting the likelihood of disease in association with exposure to a foreign substance is referred to as risk assessment. The process is now generally considered to include the identification of a potential disease risk due to exposure to a hazard, determination of sources and magnitudes of exposure to the agent, estimation of the relationship between the potential risk or severity of disease and the dose of the agent, and the integration of this information into estimates of potential risk associated with various exposure conditions (NRC, 1984a). The quantitative estimation of potential human health risk associated with exposure to a chemical generally involves an assessment of biomedical data to determine the likelihood that the chemical can produce human disease and an assessment of the dose of the chemical. Most human risk assessment today is based on estimates of external exposure (in parts per million or mg/m^3), which can be used to estimate dose (commonly expressed in mg of chemical inhaled, ingested, or absorbed per day per kg of body weight or per unit of surface area) (Anderson, 1982). A dose can be multiplied by the estimated potency of the chemical to estimate the risk per unit dose. A major limitation of that approach is that exposure data are generally imprecise and contribute considerably to uncertainty in the risk estimate. Tissue concentration, which can provide quantitative data on internal dose or biologically effective dose of a chemical or on the resulting biologic effects, can in theory introduce much greater precision into the risk-assessment process than the use of crude exposure data. Concentrations of environmental chemicals in human tissues can be used to assess the likelihood that disease will result from chemical exposures. The monitoring of human tissues for toxic substances is concerned primarily with measurement of exposure, although some information is relevant to biologic effects.

For some diseases and exposure conditions, threshold doses have traditionally been assumed—i.e., doses below which discernible effects are not elicited. For example, thresholds may be produced by reserve biologic capacity or by repair mechanisms that are fully effective at low doses. In such situations, the doses of concern are the threshold dose and larger doses. However, the occurrence of effects might not be a linear function of dose. For example, a threshold may vary from person to person. Carcinogens and some chemicals that augment existing disease processes are commonly assumed to carry some risk (perhaps very small) at even the smallest exposure (Peto, 1978). Even if a chemical is present below a threshold dose when acting in isolation, the combination of endogenous and exogenous factors already present may surpass a threshold dose.

A change or difference in the concentration of a toxic chemical in a particular tissue is generally assumed to reflect a difference in risk. For example, if tissue concentrations of DDT decrease, the potential risk of an adverse health outcome of DDT exposure is assumed to decrease. Tissue concentrations of a substance can provide an integrated assessment of the exposure to the substance by all routes, such as dermal exposure, inhalation, and ingestion.

In analysis of tissues for concentrations of chemicals, compositing of specimens (i.e., pooling of tissue specimens and analysis of the pooled sample) is a satisfactory technique if one needs only the average potential risk for the population represented by the composite. Compositing usually is not a good choice, except under the most extreme pressure. When n equaled-sized specimens are composited and found to have mean concentration $\bar{x}$, every one of the original specimens must have concentrations $\leq n \cdot \bar{x}$. If the distribution of potential risk among individuals is of interest, as is often the case, individual specimens are required.

Where no trends in average tissue concentration are noted for a population, then data over that period of time can be cumulated to improve estimates of potential risk. (Estimates might also be derived when there is a trend, but the statistical procedures are more complex). Establishment of baseline concentrations of chemicals in tissues can be useful in determining whether tissue concentrations, and hence potential risks for various populations, are abnormal.

Description of dose gives information on the intensity of exposure. Dose has been defined in many different ways, but fundamentally refers to an integrated assessment of environmental exposure (frequently termed "external dose") or of the quantity of the chemical in the body (termed "internal dose"). Relevant information includes the amount (e.g., mg/kg body weight), duration, time pattern (e.g., continuous or intermittent), and route of exposure. For environmental chemicals, such information rarely is well established. Consequently, many studies of environmental chemicals use biologic monitoring of human tissues.

It is important to distinguish between the current use of human-tissue monitoring data and their potential use in risk assessment. Data on tissue concentrations of xenobiotics constitute measures of internal dose and so can help to identify a hazard (i.e., a qualitative risk) for further surveillance or followup studies. Tissue concentrations of environmental agents in different populations can be compared (e.g., urban vs. rural and children vs. adults). Tissue concentrations can also reveal exposure trends. However, those measures of internal dose generally are not themselves directly usable for predicting potential health effects. Estimates of potency of toxic agents are generally expressed as risk per unit of whole-body dose, so the supporting estimates of exposure currently used to calculate absolute risk in risk assessment are based on external (usually environmental) concentrations. This may change as data and experience accumulate for the estimation of health effects from internal doses.

Mechanistic models relating tissue concentrations and environmental concentrations would require information on variations of contamination over time, routes of exposure, biologic factors (bioavailability and rates of uptake, distribution, metabolism, and elimination). Those kinds of information are rarely available, and estimates of exposure based on tissue concentrations typically use simplified models that incorporate limiting assumptions. The implication regarding the utility of tissue monitoring data is that their applications depend heavily on input from other fields, such as toxicology and exposure assessment, and that precise estimates of exposure based on tissue concentrations (or vice versa) are ordinarily beyond the reach of present models. For example, it is not possible to "back-calculate" the history of exposure from a single blood sample. Thus, the use of tissue monitoring data to estimate potential risks with the conventional approach is not always accurate. However, it is often possible to estimate past exposures, given limiting assumptions (i.e., a steady state) and groups of samples (in which some random effects can be averaged out). This approach is useful in comparing exposures and potential risks among population groups.

Viewed another way, tissue concentrations can sometimes be used as a direct measure of internal dose, and internal dose can sometimes replace estimated exposure as the dose variable in risk assessment. But, relating risk directly to internal dose requires more data. These relationships hold the promise of an exposure or dose measure that will yield a prediction of effect (i.e., potential risk) that is physiologically relevant. For reasons discussed above, the promise cannot yet be realized.

Biologic Markers

Sophisticated laboratory techniques developed during the last decade can now detect exposures to pollutants at very low concentrations, and can assess their behavior, fate, and effect at the cellular or molecular level. The methods have stimulated interest in the use of biologic markers in environmental research, particularly in the study of the somatic effects of exposure to environmental carcinogens and mutagens.

A biologic marker is an alteration that occurs at the biochemical, cellular, or molecular level on the continuum between exposure and disease and that can be measured with assays of body fluids, cells, or tissues. Biologic markers are discussed in depth by Perera and Weinstein, 1982; Harris et al., 1987; Perera, 1987; Schulte, 1987; Hulka and Wilcosky, 1988; and NRC, 1989. For an exposure to result in an adverse health effect, it must generate a chain of events, each of which theoretically can be reflected in a biologic marker.

Conventionally, three broad categories of markers have been distinguished: markers of exposure or dose, markers of effect, and markers of susceptibility. A distinction can be drawn between a marker of exposure (which gives qualitative information as to whether an organism has been exposed) and a marker of dose (which provides a quantitative measure.) Markers of internal dose directly reflect the parent substance, its metabolites, or its derivatives in cells, tissues, or body fluids. Sensitive physicochemical and immunologic methods can detect and measure very low concentrations of foreign substances in the body. Exhaled air, blood, and urine are ordinarily used, but other body fluids—such as breast milk, semen, and adipose tissue—have also been used. Each biologic material has its own relevance to both exposure and health outcome, and the differences affect interpretation of results.

Markers of Exposure or Dose

Monitoring for these markers is based on direct measurement of concentrations of the parent compound in cells, tissues or body fluids. Physical, chemical, and immunologic methods can now detect and quantify very low concentrations of xenobiotic substances in the body. Biologic markers of internal dose can be characterized according to their chemical-specificity/selectivity, with selective markers representing measures of pollutants or their metabolites detected in biologic media. Examples of internal dosimeters include blood levels of styrene, pesticides, and metals; exhaled volatile organic chemicals (VOCs); concentrations of polychlorinated biphenyls (PCBs), DDE (a metabolite of DDT), and TCDD (dioxin) in adipose tissue; and urinary mandelic acid resulting from styrene exposure. Various nonselective markers such as urinary excretion of thioethers and the mutagenicity of urine and other body fluids have also been assessed in humans, the latter fairly widely. Disadvantages of internal dosimeters are that, while the laboratory methods may be highly sensitive, in the absence of bioaccumulation, markers can reflect only recent exposure. This can be a limitation if exposure has been interrupted or if past exposures must be estimated. In addition, internal dosimeters do not reflect critical interactions with macromolecules in target cells (Lucier and Thompson, 1987). Biologic monitoring of dose may also be based on measurement of a metabolite of the environmental chemical or of a compound produced by cells as a result of interference of the environmental chemical in the normal metabolic process. For example, biologic monitoring for the effects of lead exposure have frequently been based on measurement of the concentration of various intermediates of heme metabolism, which is disrupted by lead. Because estimates of the amount of the chemical present in the entire body are rarely possible, and may not be as useful, a particular tissue is typically chosen as an indicator of body burdens or stores of the chemical. The specific tissue chosen in biologic monitoring may be selected because this tissue accumulates the chemical (e.g., adipose tissue for some pesticides; bone for lead or strontium) or is affected by the chemical. Measurements can also be made on exhaled air, blood, and urine, and other body fluids such as breast milk or semen have sometimes been used. Each of those biologic media has a different relationship (e.g., temporal) to both past exposure and future health outcomes. Those differences can markedly affect interpretation of results.

Markers of Effect

Biologic monitoring for the effect of an environmental chemical may be based on a change at the tissue, cellular, or molecular level in response to exposure to the chemical. Such changes may be biologic indicators that do not impair function, but serve as indicators of exposure. Generally, the effect is the biologic response (i.e., the reaction or the response of the person) following exposure to the chemical. This reaction may vary with respect to the quality, strength, onset, and duration of exposure to the chemical. A large number of factors influence this reaction. Biologic monitoring that is based on the effect produced by the chemical frequently has far greater implications than does direct monitoring of tissues for the chemical of interest. In some cases, the biologic indicator may be useful as a surrogate for measurement of the chemical or physical agent per se. The biologic indicator may be of particular value if data are available to determine the relationship between external exposure, the biologic indicator, and the toxic effect(s) of the chemical (Mahaffey, 1987).

Effect monitoring for chemical exposures can be based on assessment of a variety of biochemical indicators, including enzyme activities that can be measured either in target organs or in body fluids such as blood and urine. Examples include the inhibition of pseudocholinesterase in plasma, elevation of glutaryltranspeptidase in serum or urine, increased excretion of urinary or fecal porphyrins, D-glutaryltranspeptidase in serum or urine, and DNA adducts in urine. Other effect indicators include changes in genetic material, as identified by chromosomal aberrations, effects on the immune system (e.g., surface markers and induction of lymphocyte subpopulations), and the endocrine system. Markers in this category also include DNA and hemoglobin adducts in peripheral blood and other cells and tissues (e.g., lung macrophages, buccal mucosa, bone marrow, placental tissue, lung tissue).

To estimate risk of disease from human tissue concentrations requires knowing the relationship between concentration and the risk or extent of the development of the disease process—that is, the association of internal dose and response. This information generally does not exist and often requires extensive biokinetic research. However, relative risks (e.g., male vs. female, urban vs. rural, current vs. past) can be estimated if the incidence of disease is proportional to internal dose as measured by tissue and/or blood levels. If risk is proportional to blood or tissue levels of a chemical, then relative risks can be estimated. Uncertainties exist in relating tissue concentrations to dose and large uncertainties exist in relating risk to dose (Allen et al., 1988). Uncertainties in interspecies extrapolation of risk estimates among rodents are up to a factor of 10 to 100 for carcinogens (Gaylor and Chen, 1986). Hence, smaller uncertainties in the estimates of tissue concentrations or dose may not negate their usefulness for risk assessment.

Markers of Biologically Effective Dose

Markers of biologically effective dose measure the amount of a pollutant or its metabolites that has interacted with cellular macromolecules at a target site or an established surrogate. Measures of biologically effective dose include DNA adducts and hemoglobin adducts in a range of cells and tissues, including peripheral blood, bone marrow, lung macrophages, buccal mucosa, placental tissue, and lung tissue. Many carcinogens are metabolically activated to electrophilic metabolites that covalently bind to DNA. Adducts on critical sites of DNA, if unrepaired, can cause gene mutation, which is a critical initial step in the multistage carcinogenic process. Several methods to detect DNA-chemical adducts in lymphocytes and target tissues are currently available, including radio- and enzyme-linked immunoassays utilizing polyclonal or monoclonal antibodies, ^{32}P post-labeling, and synchronous fluorescence spectrophotometry (Santella, 1988).

For example, antibodies were used to detect the presence of polycyclic aromatic hydrocarbon-DNA (PAH-DNA) adducts in lung tissue and peripheral white blood cells (WBC) from lung cancer patients and controls. Antibody methods have also been used to analyze white blood cells and other tissues of individuals exposed to polycyclic aromatic hydrocarbons (PAHs) in cigarette smoke and in occupational settings (Shamsuddin et al., 1985; Haugen et al.,

1986; Harris et al., 1987; Perera et al., 1988b; Hemminki et al., 1990). Examples of antibodies available to assess formation of DNA adducts in humans include those to aflatoxin B_1, alkylating agents, 4-aminobiphenyl, benzo[a]-pyrene (BP) and PAHs, cisplatinum, and 8-methoxypsoralen. Immunoassays can detect adduct concentration as low as one adduct per 10^8 nucleotides.

The ^{32}P post-labeling technique is even more sensitive, as it can measure one adduct per 10^9-10^{10} nucleotides (Randerath et al., 1988). It has been applied to the measurement of adducts formed by various alkylating and methylating agents, aromatic compounds (e.g., BP/PAH) and cigarette smoke constituents (Dunn and Stich, 1986; Everson et al., 1986; Phillips et al., 1988; Hemminki et al., 1990). The method produces images that are considered idiosyncratic "fingerprints" of exposure.

A third approach, synchronous fluorescence spectrophotometry, has recently been applied to human samples with a reported sensitivity of one BP adduct per 10^7 nucleotides (Vahakangás et al., 1985). High pressure liquid chromatography (HPLC) and fluorescence spectroscopy have been used to detect excised carcinogen-DNA adducts in urine (Autrup et al., 1983).

Assays that measure carcinogen-protein adducts, especially the binding of metabolites with hemoglobin, are in some cases a good surrogate for DNA-adduct measurements. Methods available for measuring these adducts include immunoassays, gas chromatography-mass spectrometry (GC-MS), ion-exchange amino acid analysis, and negative chemical ionization mass spectrometry (NCIMS). This last method has been successfully applied to the quantitation of hemoglobin adducts formed by ethylene oxide (Osterman-Golkar et al., 1984) and 4-aminobiphenyl-hemoglobin (4-ABP) (Bryant et al., 1987). Because of the 3-month lifespan of hemoglobin, these assays reflect relatively recent exposures, while DNA adducts can assess exposure integrated over a much longer period.

Markers of Early Biologic Effect

Whereas markers of biologically effective dose indicate an interaction with critical macromolecules that might potentially result in disease, these markers may also be repaired, or otherwise "lost." In contrast, markers of early biologic effect indicate the occurrence of irreversible toxic interactions either at the target or an analogous site.

Markers of early biologic effect indicate an event resulting from a toxic interaction of a xenobiotic substance, at either the target or an analogous site, which is known or believed to be a step in the pathogenesis of disease or to be qualitatively or quantitatively correlated with the disease process. Like markers of biologically effective dose, markers of early biological effect provide integrated "black box" measurements of the net result of all the biological processes that occur when the body is exposed to a particular pollutant or pollutants (Hoel et al., 1983). As understanding of the range and complexity of the mechanisms of action of chemicals expands, such concepts may require revision in response to more fundamental mechanistic information on how toxicity occurs at a subcellular level. These include pharmacokinetic events occurring on the cellular or systemic level such as absorption, metabolism, detoxification and elimination, as well as macromolecular processes such as binding, repair, and immune response. An irreversible effect can be due to direct attack by the chemical (e.g., genotoxic effect, allergic effect, cytotoxic effect), to an indirect reaction of an organ which is not directly attacked (e.g., central nervous system damage following carbon monoxide intoxication), or to inappropriate tissue repair (Mahaffey, 1987).

Cytogenetic techniques provide a direct, though nonspecific, method of assessing changes at the chromosomal level. These changes include alterations in chromosome number, structural changes such as breakage and rearrangement, and exchanges between reciprocal portions of a single chromosome (sister chromatid exchanges or SCE). Elevated frequencies of chromosomal aberrations and/or SCEs have been observed in persons exposed to ionizing radiation or to a variety of chemicals including vinyl chloride, styrene, ethylene oxide, and organophosphates (Evans, 1982; Vainio et al., 1984). Although SCEs are a biologic effect, the significance of this increase in relation to disease outcome in unclear. Micronuclei (MN)—fragments of nuclear material left in the cytoplasm following replication—are generally considered to indicate prior chromosomal aberrations.

Another approach to assessing genetic effects involves measurement of single-strand breaks in lymphocyte DNA (Walles et al., 1988). In addition, DNA hyperploidy measured in exfoliated bladder and lung cells has been shown to be a biologic marker of response to carcinogens (Hemstreet et al., 1988).

An important new marker is the activated oncogene and its protein products. During oncogenesis, a normal segment of DNA (termed a proto-oncogene) is activated to a form that causes cells to become malignant. Activation can occur through several mechanisms including gene mutation, chromosome breaks, and rearrangements. Activation of oncogenes or their protein products can be measured by complex immunoblotting techniques or by the polymerase chain reaction (PCR) technique. More studies are needed to understand the precise role of oncogenes and their protein products in tumorigenesis and, in particular, the nature of barriers to malignant transformation (Weinberg, 1989).

Markers of Susceptibility

Generally, the effect of interest is the biological response following exposure to the chemical. This response may vary with respect to the quality, strength, time of onset, and duration of exposure to the chemical. A large number of factors influence this reaction, including age, genetic makeup, gender, developmental stage, physical activity, and normal physiological states such as pregnancy and lactation. Susceptibility to the effects of a chemical results from differences in transfer of the toxic chemical from the external dose to the internal dose, the biokinetics of distribution among and within tissues, and the responsiveness of the target tissue. Not all tissues are equally responsive to the potential effects of exposure to chemical contaminants. For example, the severe neurological effects of exposure to methylmercury during gestation are highly dependent on the developmental stage of the nervous system when exposures occur.

Markers of susceptibility reflect inherent or acquired differences affecting an individual's response to exposure. These differences can serve as effect modifiers and thereby increase or decrease risk at any point on the continuum between exposure and the emergence of symptomatic disease. Markers of susceptibility may indicate the presence of inherited genetic factors that affect the individual or a population of which he/she is a part. They may also reflect certain host factors, such as lifestyle, activity patterns, prior exposures to environmental toxicants, or nutritional status.

For example, nutritional status and individual variability affect the cytochrome P-450 system for metabolism of xenobiotics, and individual variability in cytochrome P-450 metabolism may explain differences in lung cancer risk, although results of various studies have been conflicting (Karki et al., 1987). Enzyme activity as measured by metabolism on indicator drugs such as debrisoquin and antipyrine may be useful in identifying genetic polymorphisms that modulate the effects of exposure (Conney, 1982; Harris et al., 1987). Nutrients influence the rate of cellular metabolism that may increase or decrease toxicity of an environmental chemical. For example, cytochrome P-450 concentration in liver microsomes can be lowered by protein deficient diets (Marshall and McLean, 1969). Concentrations of various cytochromes, including P-450, are decreased in the liver, kidney and adrenals of guinea pigs following ascorbic acid deprivation (Degkwitz et al., 1975).

Nutritional status may also influence the percent and rate of absorption of the toxicant, and nutrients can influence the quantity of the toxicant that reaches a critical target tissue by sequestering the toxic compound into body depots. In such depots as adipose tissue or bone mineral, the intensity of exposure to the compound in organs such as brain, liver, or kidney is relatively less than would occur if the compound were more evenly deposited. On the other hand, exposure to target organs may be greatly improved and/or prolonged by storage in tissue depots, if these depots are mobilized.

Nutrients and toxicants may affect the same biological markers. For example, both lead toxicity and iron deficiency result in impairment of hematopoiesis, and increases in the concentration of erythrocyte protoporphyrin has been widely used as a marker for both iron deficiency and lead toxicity (Mahaffey and Annest, 1982). A potentially serious complication is that the effect of both conditions produces a greater than additive increase in impaired heme biosynthesis (Mahaffey and Annest, 1982).

Age of the subject can have a marked effect on a biologic marker. An important example is exposure during gestation. The toxicity of methylmercury and lead differ greatly when exposures occur during gestation and the first few years of life when contrasted to their effects among adults. A particular concentration of lead in bone must be considered relative to age. Adults store approximately 90 to 95% of their total body burden of lead in bone. Young children, by contrast, have only about 70% of their total body burden of lead in bone, with a far greater fraction of the total body burden present in brain and other tissues.

Markers of susceptibility may also indicate a pre-existing disease condition that could increase an individual's risk. The presence of certain rare hereditary diseases (e.g., ataxia telangiectasia, Fanconi's anemia, and Bloom's syndrome) may indicate heightened susceptibility to potentially genotoxic ambient exposures (Carrano and Natarajan, 1988). Pre-existing conditions may either affect an individual's metabolic status or establish sites within the DNA where initiating events such as point mutations or translocations are more likely to occur. Restriction enzyme DNA fragment-length analysis of genetic polymorphism (RFLP) is a new approach using recombinant techniques to detect DNA conformations associated with genetic predisposition to cancers such as retinoblastoma (Francomano and Kazazian, 1986).

Summary

Although many biologic markers are in the validation stage, they have considerable potential in risk assessment and environmental epidemiology (reviewed in Perera, 1987; Hulka and Wilcosky, 1989). For example, as the above discussion has demonstrated, use of markers can improve exposure assessments and can provide timely identification of groups and individuals at potentially elevated risk of disease. Biologic markers can improve understanding of the mechanisms of disease causation and, in addition, serve as a bridge between studies of experimental animals and humans experiencing the same exposure to chemicals. Thus, biologic markers can improve risk extrapolation between species and can provide a valuable tool for health risk assessments through human tissue monitoring.

Priorities for research, testing, and regulation are based on complex considerations that include estimates of risk attached to environmental contaminants—i.e., the likelihood that human disease will arise from exposure. If a precise relationship between risk of disease associated with chemicals and tissue concentrations of the chemicals can be established, the tissue concentrations corresponding to an allowable magnitude of risk can be estimated. The selection of tissues and chemicals for monitoring in the NHATS has so far been based on persistent chemicals, such as chlorinated hydrocarbons, that accumulate in adipose tissue. In a future national monitoring network, samples might also be analyzed for selected markers of biologically effective dose and effect to provide information on possible previous exposures.

CHOICE OF TISSUES TO MONITOR

Background

A monitoring program to survey xenobiotic chemicals in human tissues should provide an estimate of chemical concentrations in the tissues selected, identify the tissues and toxic effects relevant to each chemical, and identify potential risks. Those objectives are seldom achieved in the laboratory, and they are even less likely to be achieved in a survey of tissues from the general population. It is not feasible to study a broad range of tissues in a general population sample. One must instead try to identify tissues that most nearly account for the body burden of most of the chemicals of concern.

The concentration of any xenobiotic in tissues of humans or other animals depends on several factors. The most obvious and important are the magnitude of the exposure that led to the presence of the substance in tissue and the degree of persistence of the substance in tissue. Presence indicates some exposure, but that exposure might have occurred in the workplace, in the home, or in any of several other environments, and it might have occurred at any earlier time. Persistence is a property of both substance and the specific tissue. Most compounds to which humans are exposed, including environmental contaminants, are rapidly cleared through the conventional routes of metabolism and elimination. Some notable exceptions are 2,2,-bis(p-chlorophenyl)-1,1,1-trichloroethane (DDT), other halogenated insecticides, and certain industrial chemicals such as polychlorinated biphenyls (PCBs), which are not readily cleared but accumulate in tissues in proportion to overall exposure. Those chemicals also accumulate and concentrate in the food chain and may reach toxic levels. This fact is of great concern to those responsible for the protection of the health of our population and has been a driving force behind the NHATS program.

When it was initiated, the NHATS was designed to survey concentrations of pesticides in tissues of the general population. The pesticides of greatest concern were halogenated hydrocarbons, primarily halogenated aromatic hydrocarbons. Those compounds are generally highly lipophilic and slowly metabolized—both properties that result in their accumulation in the environment and in the tissues of higher animals, including humans. The lipophilicity results in their moving from blood into tissues with a high fat content, particularly adipose tissue, where they concentrate with continued exposure. Therefore, the choice of adipose tissue to monitor for pesticides addressed the original purpose of the survey. However, the goals and objectives of the program have expanded greatly since its inception, and no tissue can be best for all purposes. Given the advances in analytic chemistry, the increased sensitivity of equipment, and the discontinuation of use of most halogenated aromatic pesticides and many halogenated aromatic industrial chemicals, including PCBs, it seems reasonable to reconsider which tissues to monitor.

One critical element in the choice is tissue-to-blood ratios of chemicals of concern. Blood is the common intermediary among all tissues. As it transports nutrients, oxygen, and wastes to and from tissues, blood maintains intimate contact with each other tissue and thereby in effect keeps all tissues in contact with one another. That contact is critical to the maintenance of homeostasis by hormones and the other mechanisms that regulate biologic processes. Most xenobiotic chemicals in the body cross cell membranes by passive diffusion, so they readily partition, up to the tissue-to-blood ratios, from exposed tissues to blood and from blood to other tissues. The result of the various tissue-to-blood ratios is a dynamic equilibrium of xenobiotic concentrations between tissues, as is shown schematically for a pharmacokinetic model in Figure 3-1. At equilibrium, or steady state, the proportions of a chemical stored in the various tissue compartments depends on the affinities of the chemical for the tissues in question and the tissue volumes. Tissue-to-blood ratios for more polar compounds often are near unity (Table 3-1); the choice of a tissue to sample depends heavily on the properties of the chemical of interest, including lipophilicity, solubility, and ease of metabolic degradation. Tissues reach equilibrium at rates that depend in large part on the concentration of the chemical in blood, the rate of blood flow to the major tissue depots, and the tissue volumes. Final tissue-to-blood ratios reflect the affinities of the chemical for the individual tissues; the ratio for lipid-rich tissues is very high for highly lipophilic chemicals.

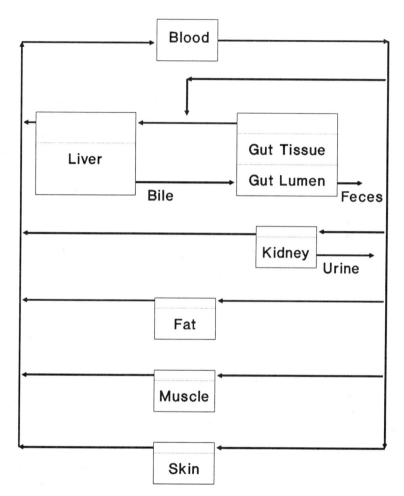

FIGURE 3-1 Schematic diagram for a typical whole-body physiological pharmacokinetic model. Source: Bungay et al., 1979.

TABLE 3-1 Examples of Tissue-to-Blood Ratios for Various Classes of Chemicals[a]

Chemical	Adipose Tissue	Liver	Kidney	Muscle	References
Halogenated aromatic compounds					
DDT	184-792	2.4-4.4	2.7-3.9	—	Wooley and Talens, 1971
DDE	52-412	1.1-5.6	1.1-2.3	—	Wooley and Talens, 1971
TCDD	89-135	121-280	—	—	Birnbaum, 1986
4-Chlorobiphenyl	30	1	—	1	Lutz et al, 1977
4,4'-Dichlorobiphenyl	70	3	—	2	Lutz et al, 1977
2,4,5,2',4',5'-Hexachlorobiphenyl	400	12	—	4	Lutz et al, 1977
Halogenated aliphatic compounds					
1,2,3-Trichloropropane	15.0	3.8	2.0	0.6	Volp et al, 1984
1,1,1-Trichloroethane	16.0	2.7	3.2	1.0	Matthews, 1988
Bromodichloromethane	1.0	14.3	6.5	0.6	Matthews et al., 1990
Aromatic amine compounds					
p-Nitroaniline	7.0	0.9	0.4	0.8	Chopade and Matthews, 1984
Benzidine	†	†	†	†	Lynn et al, 1984
p-Phenylenediamine	0.1	3.0	—	0.7	Ioannou and Matthews, 1985

Miscellaneous compounds

2-Butoxyethanol	‡	‡	‡	‡	Ghanayem et al., 1989
Chlorendic acid	—	5.7	1.4	—	Decad and Fields, 1982
Methyl carbonate	0.17	0.9	—	1.0	Ioannou et al., 1988
Ethyl carbonate	0.02	0.9	0.8	0.6	Nomeir et al., 1989
Benzene	—	1.1	—	—	Sabourin et al., 1989
Theophylline	0.1	1.2	1.2	1.1	Shum and Jusko, 1987
Tris(1,3-dicholoro-2-propyl) phosphate	1.2	1.7	1.4	0.4	Nomeir et al., 1981

[a]Tissue-to-blood ratios for parent compounds in rats or mice determined with radiolabeled chemicals and various analytic procedures.

†Not present in quantifiable amounts 2 h after dosing.

‡Not detectable in blood 1 h after dosing.

The passive diffusion of chemicals across cell membranes accounts for the fact that, if a chemical can be eliminated from any tissue, the constant equilibration between blood and all tissues will result in its elimination from the body. Conversely, if a chemical cannot be readily eliminated, a decrease in its concentration in one tissue will increase its concentrations in other tissues. That relationship accounts for the observation that rats exposed to DDT at nontoxic concentrations for 3 months died of DDT intoxication when placed on a restricted diet that reduced their total body fat and consequently increased concentrations (Kimbrough, 1982).

Blood

If blood-to-tissue ratios are known, determination of the blood concentration of a chemical can provide accurate estimates of other tissue concentrations. The characteristics that determine chemical partitions between tissues are the same in humans and animals, so blood-to-tissue ratios in rats and mice, such as those in Table 3-1, yield good approximations for humans. Furthermore, they are available for many compounds and so could be easily developed to support surveys of human exposure and tissue concentrations. Thus, assay of blood samples should permit accurate estimation of tissue burdens of many chemicals of interest.

Detection of a chemical in blood or another tissue does not, however, provide insight into the time or magnitude of exposure. Analysis of tissues for chemical content can confirm exposure or lack of exposure, but only serial samples from a single person or many samples from a population over a long period can yield an accurate estimate of exposure or persistence. Blood concentrations of many chemicals change greatly within hours or days of exposure. Some chemicals, such as benzidine and the glycol ether 2-butoxy-ethanol (Table 3-1), are no longer detectable in blood within a few hours of exposure and some (e.g., formaldehyde) disappear even faster; they are completely and rapidly cleared from other tissues as well, so no tissue has evidence of earlier exposure.

A disadvantage of sampling blood is that it often contains lower concentrations of the most persistent xenobiotic chemicals than any other available tissue, so assay sensitivity is low. In contrast, although blood and adipose tissue can differ considerably in their concentrations of some highly lipophilic chemicals (Table 3-1), the difference is relatively minor for most other classes of chemicals. With improving analytic methods, this disadvantage is not as great as it was when the NHATS program was initiated, and it should become even less important as methods continue to improve. Another important issue that must be addressed for blood collection is the safe and proper handling and storage of samples. The potential health consequences of careless or improper handling of blood can expose laboratory workers to viruses, such as hepatitis B and AIDS. Individuals involved in collection, analysis, and storage of blood must be given careful direction and supervision.

Numerous chemicals can be detected in the breath of humans (Wallace et al., 1986; Gordon et al., 1988). The chemicals so detected are most often volatile chemicals encountered in the home or workplace. Breath analysis is in effect a blood analysis, because there is a constant passive equilibration of chemicals between blood and air in the lungs, just as there is between blood and other tissues. The concentration of a chemical in exhaled air will be proportional to its volatility and its concentration in blood and inversely proportional to its solubility in blood. However, most chemicals are insufficiently volatile to permit detection in breath, and the ones that are sufficiently volatile are usually so depleted as to be undetectable within hours, or at most a day or so, of exposure. But breath analysis can yield estimates of relatively constant exposure, as sometimes occurs in the home or workplace.

An important advantage for blood collection, in contrast to collection of adipose samples from cadavers, is that samples to be obtained from living persons, so that interviews can be conducted with sample donors to obtain demographic and environmental information, which permits examination of causal relationships and risk factors. For example, interviews can yield data on the following:

• Geographic location, including current residence and length of stay in previous residences. This information could be used for analysis by region, for urban-rural comparison, for separate analyses of persons living near heavy industry or chemical or petrochemical refineries, etc.
• Demographic information, such as sex, age, race, and ethnicity. Separate analyses of such groups are of interest.
• Information on occupation and industry, particularly whether employment is in a chemical plant or refinery, on a farm using chemical pesticides, etc.
• Other environmental data, such as type of drinking water used (private well water or community system).

The number of samples in a single year might not be large enough to support detailed analysis of the relationship between the information obtained during interviews and contaminant concentrations in blood, but data from 2 years or more can be combined. With the larger number of samples, relationships among tissue concentrations, demographic data, and occupational and environmental exposures can be studied with more precision.

In developing a monitoring system that relies on large samples of blood, multiple factors must be evaluated, including the need for special storage methods and containers. Many such issues may have been studied previously by other groups (e.g., the Red Cross); however, it is unlikely that all storage methods and containers have been evaluated to identify factors that might influence the measurements of environmentally important chemicals. Included in the questions that should be studied with great care are to what extent freezing/thawing will change concentrations of volatile contaminants; whether and how cells should be separated from plasma or serum to minimize analytical effects; and how samples should be stored before analysis, including temporal effects on volatile chemical concentrations.

Adipose Tissue

Adipose tissue contains the highest concentrations of some of the most persistent chemicals in the environment, including halogenated aromatic hydrocarbons. Those concentrations can be 100 or even 1,000 times greater than concentrations in blood or other lean tissues. Assay of adipose tissue can thus greatly enhance sensitivity. However, tissue-to-blood ratios of adipose tissue other tissues are much smaller for most chemicals (Table 3-1). The use of adipose tissue is also complicated by the difficulty of obtaining samples of fat. It is because of that difficulty that the NHATS program obtains approximately 80% of its samples from autopsies. Satisfactory samples can be obtained from autopsy, but reliance on this source limits both the representativeness and the numbers of samples available.

Whenever possible, samples of adipose tissue should be taken from persons who were known not to have died from long-term illnesses or wasting diseases (e.g., cancer or AIDS). Furthermore, samples should be taken from the same anatomic site in each individual from the sampled population, in order to minimize the inherent variability among sampled sites and individuals.

Lean Tissues

Of the lean tissues, liver and kidney usually contain the highest concentrations of organic chemicals. Muscle has the largest tissue volume and might contain a major portion of the body burden of a chemical, but concentrations in muscle are usually similar to those in blood (Table 3-1).

The kidneys are major excretory organs and can contain transient high concentrations of some chemicals. Other chemicals can be concentrated and retained by the kidneys as they recover water, minerals, and other essential substances from the glomerular filtrate.

The liver plays a dual role of xenobiotic degradation and excretion in bile. Most of the metabolism of foreign chemicals occurs in the liver, and some can be concentrated in the liver during metabolic degradation. Some chemicals, such as the tetrachlorodibenzo-p-dioxins (TCDDs), seem to have an affinity for the liver, even though they are not readily degraded (Birnbaum, 1986). Others, such as chlordecone (Kepone), are concentrated in the liver as a result of hepatic excretion in bile, reabsorption from the intestine, and return to the liver in enterohepatic recirculation (Bungay et al., 1979).

Lean tissues are primary depots for heavy metals (Comar and Bronner, 1964). Cadmium and mercury are concentrated in the kidneys, and cadmium in the liver. Lead is concentrated in the liver, but more concentrated in bone. Each lean tissue can be the preferred tissue to assay for one or more chemicals. Lean tissues are generally available for assay only from autopsies; however, once a collection system is established for collecting adipose tissues at autopsy, then other solid (lean) tissues, such as liver and kidney, can be collected without much additional expense or effort.

Biologic Fluids

Just as the equilibration of chemicals between blood and other tissues results in a dynamic equilibrium among all tissues, there is an equilibration between tissues and the fluids that they secrete or excrete. Almost all biologic fluids (urine, sweat, saliva, milk, etc.) contain chemicals at concentrations proportional to those in the tissues in which they originated. But most (milk is an exception) are rather polar media and contain low concentrations of the persistent lipophilic chemicals that have been of most concern.

Milk contains lipid, as well as protein and water, and has been an important indicator of exposure in several incidents of environmental contamination with lipophilic compounds (Matthews, 1979). However, milk samples can be provided only by postpartum women, a highly limited and atypical segment of the population useful mostly for studies directly relevant to the fetus or nursing infant.

In studies of metabolism, urine is routinely assayed for parent substances and metabolites. Because it is polar, urine contains only low concentrations of lipid-soluble parent chemicals and has been used only sparingly to determine environmental exposures. Increased use of high-performance liquid chromatography (HPLC) and characterization of major metabolites of chemicals of interest would increase the utility of urine as an assay medium.

Other biologic fluids are equally polar, but are used less than urine, because they are harder to collect, are present in smaller volumes, and generally offer no important advantages over urine. Sweat might have the greatest potential for development of a noninvasive assay for xenobiotic chemicals in humans, because it contains substantial amounts of skin oil, as well as water and minerals. If methods for standardization could be developed, sweat might be useful for assays for many xenobiotic chemicals.

Hair

Xenobiotics have been detected in hair, feathers, and nails of numerous species. Hair is the most easily obtained human growing tissue for monitoring and has received the most attention. Several toxic elements—including selenium, mercury, and arsenic—have an affinity for hair, probably as a result of their reaction with sulfur-containing amino acids, which are more highly concentrated in hair than in other tissues. Those elements are chemically bound and thus not easily removed by washing, so their presence in hair is proportional to the magnitude of exposure to them. Hair contains mercury at approximately 300 times the concentration found in blood and is considered a good indicator of the body burden of this element. But the probability of contamination from the external environment, such as selenium-containing shampoos, makes hair a less accurate indicator of exposure to other elements (EPA, 1976).

Organic compounds are generally not chemically bound to hair, but are more commonly found in oil produced by the body and associated with hair. That is particularly true of highly lipophilic substances, such as halogenated aromatic compounds. Halogenated insecticides and other halogenated organic compounds are easily detected in human hair by simple extraction and analysis with gas chromatography (Matthews et al., 1976). Because the quantities of oil and hair, the rate of hair growth, and personal hygiene all vary greatly among individuals, analysis of hair for organic compounds in hair oil can rarely be used to derive quantitative estimates of exposure; but given its ready availability and ease of assay, hair should not be overlooked as a qualitative indicator of human exposure to a wide variety of xenobiotics.

SUMMARY AND RECOMMENDATIONS

Given the central role of chemicals in modern society, people will be exposed to chemicals. It is prudent that the general population be monitored to document magnitudes of exposure and to determine the need for and effectiveness of regulations and other measures to limit risk. An essential part of monitoring exposure of the general population is a survey of chemical concentrations in human tissues. The original NHMP program was required to concentrate on chemicals in human adipose tissue. However, the objectives of the NHMP have broadened with time, the classes of chemicals currently surveyed or proposed for survey are much more varied than those originally targeted, and analytic instruments are much more sensitive than they were 20 years ago. Therefore, a survey of adipose tissue might no longer be the best way to obtain accurate estimates of concentrations of chemicals of current interest in the tissues of the general population.

After extensive discussion of the advantages and disadvantages of the major tissue groups and biologic fluids, the committee draws the following general conclusions regarding the choice of tissues on which to base a survey of chemical residues in human tissues.

• Blood is the common intermediary among all tissues, and chemical concentrations in blood accurately reflect those in all other tissues. Blood also offers the advantages of availability by a widely accepted, relatively noninvasive technique and of being the most accessible tissue for assay. An assay of chemical concentrations in blood would permit sampling of a wider sector of the population, better comparison of exposed populations with national averages, repeat sampling of persons who have high tissue concentrations, and opportunities to follow chemical clearance with time. We recommend that any new program to assay chemical concentrations in tissues of the U.S. population be based primarily on analysis of blood.

• Adipose tissue contains the highest concentrations of some of the most persistent chemicals to which humans are exposed, primarily halogenated hydrocarbons. Analysis of adipose tissue would provide the most sensitive assay of tissue concentrations of those chemicals and should be continued where feasible. Continued analysis of adipose tissue would also provide continuity with the present program, as well as confirmation that a survey based on blood also detects important tissue residues of persistent chemicals.

• Analysis of lean tissues (although not specifically recommended by the committee) and fat taken at autopsy would yield data on chemical concentrations in specific tissues that could be used to calculate tissue-to-blood ratios for estimating tissue concentrations where only blood is available.

• With additional development, analysis of hair, urine, and some other biologic tissues and fluids might provide noninvasive methods to estimate human tissue concentrations. However, for most organic compounds, those methods need additional refinement and supporting data before they can be used in general assays.

The question of which tissues to sample requires very careful consideration.

> *The present evidence leads the committee to conclude that the basis of a human tissue monitoring program should be the broad, random collection of samples of blood.*

Implementation of this recommendation includes probability sampling; data needs cannot be satisfied by existing EPA plans regarding the proposed National Blood Network to sample blood donors.

> *Blood collection should be supplemented by the continued collection of adipose tissue, in part to maintain historical continuity while new long-term series of blood measures are established and in part because some important residues are most concentrated in fat.*

As discussed elsewhere, fat samples are necessarily nonrandom and nonrepresentative, so careful study of the relation between blood concentrations and fat concentrations is needed to validate the latter.

Measurements of nonrandom adipose tissue (as in the present NHATS) will continue to be important for at least several years, although they might be replaced later with studies of the lipid fraction of blood.

> *While blood and adipose tissue are being collected, the program should devote some resources to study of the correlations between chemical measures of xenobiotics in these two tissues, so that the effects of nonrandomness in the adipose samples will be better understood and the continued contribution of the adipose samples can be properly evaluated.*

> *Whatever tissues are collected, samples should be accompanied by standardized information on demographics, illness (especially terminal illness), and known occupational or other major exposures to chemicals.*

High priority should be given to the collection of matched adipose and blood specimens for future parallel analyses. Matched specimens of fat from different anatomic regions might also be useful for methodologic studies.

4

Sampling Methods

INTRODUCTION

As is both common and appropriate in large, continuing statistical studies, the NHATS is intended to serve many purposes. EPA has described its major objectives as the detection of toxic substances in human tissue, the establishment of baselines and trend data on chemical exposures, the identification and ranking of chemicals for toxicologic testing, the identification of populations at risk and their ranking for risk reduction, and the assessment of the effects of regulatory actions. The generally accepted method of developing a data base to satisfy such objectives is to define the target population and measurement method, select a probability sample of the population, and apply the measurement methods to this sample. To move from a set of measurements to desired statistics, it is necessary to decide on the kinds of statistics that best summarize the data, to carry out computations, and to present the resulting data for a representative sample in a useful form. Sampling errors are usually calculated to give analysts and other users of the data some understanding of the effects of random variation in the data. Nonrandom variation (bias) sometimes is also discussed and assessed by such means as sensitivity analysis and (when several sources exist) reviews, including meta-analysis.

When human populations are studied, many of those steps can be carried out only imperfectly. Problems in developing a true probability sample are common, serious measurement errors might be unavoidable, and compromises in the definition of the target population are sometimes necessary. Most statisticians recognize that such problems are inevitable and accept some small or moderate departures from an ideal survey. However, when the methods deviate in important ways from accepted standards and practice, the validity of results is questionable, and the extent to which the statistics accurately reflect what is going on is uncertain. As a minimum, a statistician would like to see an analysis of departures from norms to determine whether those departures affect the quality of data produced in ways and degrees that compromise their utility.

This chapter examines the degree to which the sampling methods used in the NHATS permit inferences to be made about the total U.S. population. It deals only with sampling; measurement methods are discussed elsewhere. We describe ideal sampling practices, the compromises that are often necessary because some parts of the population are inaccessible or because of budgetary restrictions, the sampling methods used in the NHATS, and implications for drawing conclusions about the U.S. population. We also describe possible actions to improve this aspect of the NHATS without revising its general structure and consider the long-term goals of the NHATS and methods of accomplishing them.

This chapter is intended not to provide detailed guidance on designing and selecting a sample, but rather to highlight important issues. Sample design and selection are, in general, complex activities that require special expertise in sampling, not just in statistics generally. The sponsor of a new program for monitoring human tissues should engage such expertise early in the process of program development. The greatest demand for special competence will extend through the whole period of development and through the first round or two of implementation, but some need for access to sound statistical advice on sampling will continue indefinitely and might well be part of the expertise of the permanent staff. Competence in sampling should also be represented on the advisory committee that we recommend in Chapter 7.

SOME FEATURES FOR SAMPLING
IN A CONTINUING POPULATION SURVEY

An early step in the planning of any survey is to establish its primary goals in considerable detail. That should include description of the target population and, if necessary, subpopulations to be studied separately. The goals should indicate the precision desired for key statistics. For example, the survey might be required to detect whether a year-to-year change in a particular statistic had occurred and have at least an 80% probability that a 10% change over 1 year will show up as statistically significant. Or it might be desired to measure the current prevalence of some condition with an error of no more than 5%, also at the 95% confidence level. Other possibilities are that both requirements must be met simultaneously, or that one of them must apply to particular subpopulations. Precision requirements must be specified to establish the minimum sample sizes for each population group to be analyzed separately. Any subpopulations for which the population sampling fraction does not produce the sample size required for adequate precision will need to be oversampled.

In national surveys based on personal observations or interviews, the sampling is almost always based on a multistage design (Hansen et al., 1953; Cochran, 1977). Simple random sampling and other forms of single-stage sampling are generally too expensive and impractical for other reasons. The sample design in a national survey is thus usually fairly complex, and the statisticians responsible for the sample-selection methods normally go through a series of steps:

- Consider the reasons for and effects of foreseeable failures in various possible sampling schemes, such as incomplete information on some sampled persons.
- Determine what the sampling stages should be; cost, logistics, and information that is or could be made available for sample selection at each stage should be considered.
- Decide on the sources for the sampling frame at each stage. The sampling frame is the list of units from which the sample will be selected.
- Establish the modes of stratification at each stage.
- Plan for the method of sample selection at each stage. We strongly advise that probability sampling be used, but several approaches are feasible within the rubric of probability sampling, e.g., simple random sampling, systematic selection, and sampling with probability proportionate to size.
- Determine targets for sample sizes that will satisfy the precision goals of the study, including the effects of probable departures from the plan. Sample sizes at the various stages are usually calculated to provide an approximately "optimal" sample design, i.e., one that produces the lowest sampling variances for the available budget.
- Consider whether oversampling of some subpopulations is desirable. If so, efficient methods of oversampling should be developed and included in the sampling plan.
- Develop the best method of estimating the characteristics of the target total population and subpopulations from the sample. Also plan for the calculation of reasonable estimates of uncertainty in results.

Methods to carry out the various operations necessary to implement the sample design also need to be planned. Which sampling operations should be carried out by field or contract personnel, and which should be done in or from the central office under the direct supervision of agency statisticians?

How will the process of sampling be monitored, and what kinds of quality control will be imposed on sample selection? Some kinds of quality control can be effected by checking the work, and others only by comparing selected statistics that are estimated from the sample with known population data.

Statisticians generally consider that the development of a sample design should include procedures for estimating sampling errors. Most national surveys of human populations use complex sample designs, with multistage sample selection, stratification, and probabilities of selection that differ among subpopulations. The variances that are available in most of the multipurpose statistical software packages do not apply to such designs, and separate computations are needed for appropriate estimation of sampling errors. Several special software packages are available for that purpose (Flyer et al., 1989).

It is rare for statisticians to have all the data they need to design the most efficient sample. Population parameters are often only approximate. Even less is known about other factors that are normally taken into consideration, including unit costs at each stage, nonresponse rates, such logistical features as transportation, and quality control. Thus, it is advisable to conduct research on the sample design to see whether important improvements can be made. That might be a major task in initial design, but periodic review of continuing experience is usually sufficient to keep a basically sound design up to date.

Another reason for periodic review of the sample is that modifications in the objectives of the study often occur after initial survey data become available. For a continuing survey, it is sensible to see whether revisions in the sample design would support revised objectives better than the design initially chosen.

COMMONLY REQUIRED COMPROMISES
IN SAMPLING METHODS

As indicated earlier, most population surveys require some compromises from the practices discussed above. We describe here the departures from strict probability sampling that occur in many national studies and are usually accepted as having only a modest effect on the validity of resulting data. This description will serve as a crude yardstick for review of the NHATS.

Shrinking of Target Population

It is fairly common to exclude small portions of the target population from the sampling frame. They might be omitted because it would be extraordinarily expensive to include them or because they are not accessible with the data-collection methods planned. Study conclusions must, of course, be appropriately qualified or modified. Some common examples follow.

• Omission of institutional populations from the sampling frame, even though one would prefer to consider them as part of the target population. This is done in the National Health and Nutritional Examination Survey (NHANES), the National Health Interview Survey, and many other major federal surveys (Chu and Waksberg, 1988; NCHS, 1989).

• Inclusion only of persons in households that have telephones, so that data can be collected by telephone. This has become common practice for one-time surveys for both federal agencies and other sponsors and for political polls and market research.

• Restriction of the survey to the 48 states and the District of Columbia to avoid high costs of operating in Alaska and Hawaii.

• In studies of minority-group populations, restriction of the sampling frame to geographic areas that contain high concentrations of the group members. This reduces costs of screening households to identify the minority and members reduces travel costs if personal interviewing is used. The practice was followed in the Hispanic Health and Nutritional Examination Survey (Gonzalez et al., 1985).

Another type of shrinkage occurs because particular segments of the population cannot be contacted with usual survey methods. For example, in the Census Bureau's major population sample surveys, there is evidence of substantial undercoverage of homeless persons, young black males, and other minority groups. The same shortage is found in almost all population surveys. The target population thus excludes them, though not by design.

Revision of Goals to Fit Sample of Affordable Size

Standards of precision that are stated as desired in early planning stages of a survey often require sample sizes that are seriously inconsistent with the available budget. Adjustments can be made in the goals, the budget, or both. For example, one can combine data for several years, instead of analyzing each year separately, or one can broaden the demographic groups to be studied.

Acceptance of Nonresponse

Most population surveys cannot contact all sample units or get 100% cooperation from persons contacted. Nonresponse reduces the integrity of the sample, because the sampling process is not being strictly followed. Survey practitioners are resigned to the inevitability of some nonresponse, and most statisticians feel that a small amount of attrition in the sample will usually have only a minor effect on the quality of the resulting data. However, some exceptions are important and the effect of nonresponse must be considered separately for each survey. When refusal to cooperate is especially common among a particular segment of the population that differs in important ways from the population at large, even a small nonresponse rate can produce serious biases in the statistics.

Responsible survey organizations put considerable effort into attaining as high a response rate as feasible. Furthermore, statistical methods are used to adjust the resulting data to reduce, to the greatest extent possible, the potential biases introduced by nonresponse, and analytic reports explore the possible effects of nonresponse.

When knowledgeable statisticians judge the nonresponse rate to be so high as to compromise the most critical findings, the survey might have to be abandoned.

Substitution for Sample Units

In most national surveys, the samples are selected in stages. Usually, all stages except the last consist of grouping of sample units that are the subjects of the survey. For example, school children can be sampled by first selecting a sample of school districts, their subsampling schools within the sample districts, and finally choosing children within the schools. The school districts and schools, which containing groups of children, are stages in sample selection. Nonresponse sometimes occurs in the stages of the sample that consist of groups of the units of interest. In the example above, this would happen if a school district or school failed to cooperate.

Many statisticians will substitute a willing unit (e.g., school district or hospital) not initially picked for a sample for an uncooperative sample unit. An effort is made to choose the substitute within the substratum containing the uncooperative unit, on the assumption that there is reasonable homogeneity within substrata, so that the units are largely interchangeable.

Substitution is a form of adjustment for nonresponse. There is some disagreement among statisticians on whether substitution is preferable to statistical adjustment. However, all agree that stringent efforts should be made to keep nonresponse as low as practical before either substitution or some other adjustment procedure is used and that serious nonresponse at any stage should lead to reexamination of procedures used, adequacy of the data collection staffs, or other features of the survey operations that may contribute to nonresponse.

Analyses of Effects of Compromises

It is desirable for statistical agencies to expend considerable effort and resources in trying to control and understand the effects of necessary compromises on the statistics generated in their surveys. Three kinds of actions are taken:

- A substantial share of survey resources is applied to keeping nonresponse and undercoverage low. This usually requires numerous callbacks to sample units that are difficult to contact, as well as attempts to convert refusals through experienced staff members' special appeals, the efforts of survey sponsors, or in some cases peer pressure.
- Statistical adjustments are made to reduce the effects of nonresponse, undercoverage, and exclusions (by design or otherwise) from the target population. This is frequently done through poststratification and other forms of weighting the resulting data.
- Research is carried out on how a missed part of a population is likely to differ from the observed part and on how much the differences are taken into account in the statistical adjustments. Research also investigates whether better methods of adjustment exist and sometimes whether techniques are available to reduce the size of the missing portion of the desired target population.

THE NATIONAL HUMAN ADIPOSE TISSUE SURVEY SAMPLE

Goals of Study and Target Population

NHATS goals are expressed in very general terms. The committee could find no clear articulation either of its primary objectives and priorities or of the precision needed in detecting magnitudes of pollution or year-to-year changes. As a result, there is no way to determine the minimal sample sizes needed. Similarly, there is no guide to follow when choices have to be made, e.g., whether to use adipose tissue or blood, whether to choose broad-scan analysis of tissue samples, and whether to rely on composites (which make prevalence estimates impossible).

The target population is presumably the entire living U.S. population, but the subjects on whom measurements are made are an uncontrolled combination of recently deceased persons and surgical patients. There is no attempt to keep the mix of deceased and surgical patients constant from year to year. It is implicitly assumed that the distribution of pollutants is the same for deceased persons and surgical patients within the race, sex, and broad age groups for which sample quotas are designated. More important, it is assumed that substances in deceased persons and surgical patients reflect what is present in tissues of living Americans generally. The underlying assumption that long-lived contaminants in human tissue are spread fairly uniformly among recently deceased and living persons appears plausible (except perhaps for the effects of recent weight loss), but this has not been evaluated either empirically or theoretically.

The exclusion of most of the rural population from the sample frame is a serious deficiency, particularly because the detection of pesticide residues was, at least at its onset, an objective of this survey, and exposure to pesticides in rural (farm) areas can be very different from that in urban areas. It is likely that underrepresentation of the rural population leads to understatement of contamination. There is no information on urban-rural differences that would permit inferences about the effect of the rural exclusion on the statistics.

Alaska and Hawaii were also excluded from the sampling frame. Those two states currently account for only 0.7% of the U.S. population. Their exclusion should, thus, have only a trivial effect on nation-wide statistics—far less than the exclusion of the approximately 25% of the population that is outside the metropolitan sampling areas (MSAs)—but could be serious if there is a need for data specific to these areas (e.g., certain farm workers in Hawaii, or regular consumers of meat from wild animals in Alaska).

First Stage:
Metropolitan Sampling Area

Probability sampling methods were used to select the initial sample of MSAs. They were selected with probability proportionate to size, and detailed geographic stratification was used.

Unfortunately, the integrity of the sample of SMSAs has been eroded, in that 10 of the 47 sample MSAs have been replaced. For example, one SMSA was replaced when the medical examiner failed to cooperate and no large cooperating hospital in the area could be found. The nonrandom replacement of more that 20% of the MSAs could have a substantial effect on the statistics.

The substitutes for noncooperative MSAs were chosen from the same strata as the initially selected areas; they are presumably as much like the dropouts as possible. If replacements have to be made, the procedure followed in the NHATS appears to be sensible, but it should be recognized that it is the best among a set of unpleasant alternatives. There is no way to know whether refusals to cooperate are somehow related to magnitudes of contamination. In the absence of such knowledge, it is prudent to attempt to avoid substitutions.

Second Stage:
Medical Examiner or Pathologist

NHATs does not use probability sampling at the second or following stages of sampling. In MSAs containing more than one county (virtually all the large MSAs and many of the smaller ones), a single county is first selected. Specific rules for county selection are not prescribed, and the choices are left to the contractor for the project. There seems to be a preference for the first choice to be the largest county or the one with the largest city, perhaps because it will contain more hospitals and thus make finding a cooperative one easier. However, the contractor can move to other counties, if there is difficulty in getting cooperation in the initially selected counties. It is likely that there is a bias in overrepresentation of city residents, compared with suburban residents. The extent of the bias is uncertain, because many of the large hospitals draw their patients from a wider area than a specific county. That does not hold for cases brought to the county medical examiner's attention or to city- or county-operated hospitals, where the cases are much more likely to be restricted to city or county residents. It is plausible that city and suburban residents are exposed to different magnitude of pesticides and other contaminants. Sample selection methods are thus likely to influence the statistics to an unknown extent.

Within each selected county, an attempt is first made to procure the county medical examiner's agreement to cooperate. There seem to be no strong efforts to persuade reluctant examiners to change their minds: no contacts are made by high-level EPA officials, state officials, or local medical society representatives emphasizing the importance of the study; and the Midwest Research Institute (MRI) operations manual on procedures to be followed in data collection does not provide any motivation for a contractor's field representatives to attempt to convert refusals. It is implied that all potential respondents are equally acceptable. For example, the manual describes the contact with medical examiners and pathologists as aimed at determining whether "an interest exists in participating in the survey," and there is no sense of importance in having those individuals cooperate. Similarly, the letters in the recruitment package use the phrases "invitation to participate" and "we hope that you will assist." The attitude seems to be that it is not important who provides the specimens.

According to MRI, within each SMSA one or more hospitals and associated pathologists or medical examiners are selected and asked to supply adipose tissue specimens. If a medical examiner or pathologist cannot be recruited, a hospital pathologist in the same county is designated. Guidelines for choosing hospitals are more specific than those for counties. The choices are restricted to the ones that have the largest number of beds and that are full-service institutions (neonatal to geriatrics). If more than one hospital satisfies the requirements, the contractor picks one. There are no clear directions on how to do so; the manual states that "the decision is . . . then made as to which hospital would be the best choice. . . ." If the first hospital contacted does not wish to cooperate, another hospital is picked, etc. If the list of available hospitals and medical examiners is exhausted, a replacement SMSA is selected.

There is no requirement for serious efforts to motivate reluctant hospital pathologists to cooperate. The operations manual states that "the pathologist is contacted to determine if an interest exists in participating in the survey. If the answer is 'no', . . . a notation and date of nonacceptance are made . . . so the hospital is not contacted again in the near future."

With the county medical examiners and hospital pathologists chosen subjectively, it is not obvious why it makes a difference which ones agree to cooperate. However, the current system is close to self-selection. Unknown and unexpected biases are plausible under such conditions. For example, if large city hospitals are overrepresented, there is also overrepresentation of the kinds of persons who attend such hospitals, including those who are alcoholic, are homeless, are in chronic poor health, or have other characteristics that might be correlated with particular environmental exposures.

Third Stage: Specimen Donor

Each medical examiner is given a quota—by age, sex, and race—for the number of specimens to be provided in the course of a year. However, no sampling method is designated for choosing the persons or even for avoiding possible seasonal effects.

Attempts are made to attain the desired quotas for each cooperator. Field personnel are instructed to call the sample medical examiners and pathologists periodically to stimulate them to meet the quotas. Those calls appear to be ineffective: in the last few years, only about half the quotas were achieved. The quota sizes are intended to be representative the SMSA and to provide an approximately equal probability sample within the sex-age-race categories designated by the quotas. Weighting is used to adjust for departures from the quotas.

With probability sampling, the selection of first-stage units with probability proportionate to size and the use of properly calculated fixed quotas of persons within the first-stage units provide close to equal probability samples for each of the designated sex-age-race categories. The quotas have to be revised periodically to reflect changes in the distribution of the population among the first-stage units. Weighting to adjust for failures to meet quotas retains the representativeness of the sample (although it increases the sampling variance), provided that the persons on whom measurements are taken are random samples of the population within the first-stage units. However, when shortages in the quotas appear with no indication of why they occur, it becomes uncertain how well the sample cases represent the population. Because there is no describable sampling system for choosing the specimens, it is not clear whether failing to meet quotas has an appreciable effect on the representativeness of the sample.

Fourth Stage: Body Source of Specimen

In developing a monitoring system, one can choose to monitor the adipose tissue in particular parts of the body if that tissue is considered to yield an effective measure of a contaminant concentration that affects health, or one can take specimens from random regions of the body to determine the body burden. If a contaminant is uniformly distributed through the body's adipose tissue, it does not matter where the specimens are taken from. For blood samples, the assumption of uniformity seems appropriate and blood samples from any part of the peripheral circulation are reasonable.

Pathologists are given no specific instruction on the body part for specimens. It is implicitly assumed that, as is true for blood, the contaminant in question is uniformly distributed through the body.

Sampling Errors

Sampling errors have not been calculated in the NHATS. Without probability sampling, sampling errors do not have the precise mathematical properties that permit one to use the survey results to establish confidence intervals around the calculated statistics. But, sampling errors would yield a rough indication of whether apparent annual changes could be real or are likely to reflect the small size of the sample. Similarly, comparisons among demographic groups or regions produce uncertain results if there are no measures of sampling error.

IMPROVEMENTS THAT ARE POSSIBLE WITHOUT CHANGING BASIC DATA COLLECTION METHODS

The committee's review of NHATS operations led it to conclude that, although no broad sample of any human population is perfect, and compromises in sampling plans are almost always necessary, the sampling scheme used by the NHATS is far out of line with current concepts of what is acceptable.

This section discusses improvements that are possible, at a modest cost, within the present framework of the NHATS. Although the committee recommends that adipose tissue be replaced by blood as the primary human tissue monitoring program, it should be retained, with improvements, during a transition, and possibly longer. The changes described here are intended mainly to ensure that the NHATS sample adheres closely to the intended selection method, and they would increase the usefulness of the NHATS as part of a future, continuing program.

Sample Size

We have found no description of the rationale for choosing the sample sizes used in the NHATS. They might have been based on the budgets that were considered to be available for the NHATS, rather than on an analysis of the precision necessary to help EPA reach policy decisions. That is a natural approach for organizations (government agencies as well as others planning statistical studies) to take—established budgets do determine the size and scope of programs. However, when a program is as important as the NHATS, and when it is a continuing program that can be redirected and improved, decisions should be reviewed from time to time to see whether they are still applicable.

We believe that EPA should consider de novo the minimal sample size needed to satisfy reasonable goals for the NHATS. Budgets should then be estimated for that sample size (and other changes in the survey methods). We expect the budgets to exceed the annual costs of the last few years, and requests for future years' appropriations should reflect the cost of implementing the necessary changes in sample sizes. Clearly, an analysis of the needs for particular levels of precision and the sample sizes necessary to meet them will be an important justification for a new budget request.

The committee is not in a position to recommend a specific sample size, but we do suggest an approach that might help EPA to determine a sample size. EPA first must consider, more specifically than in the past, the kinds of policy-relevant questions on which NHATS data will be expected to provide guidance and, more broadly, what its responsibilities are to inform other analysts and the U.S. public on the existence of contaminants in human tissue. The questions need to be expressed in more quantitative terms than the general goals that have been stated in the past (and stated elsewhere here). The following are examples.

- For some contaminants, there should be reasonable assurance that the proportion of the U.S. population with detectable concentrations is estimated with an error rate no greater than x%.
- For other contaminants, the proportions of persons with concentrations above some specified threshold, y, should be estimated with a specified precision.
- The national mean concentrations (for example, expressed in parts per million) should be estimated with an error rate no greater than x%. In addition, the mean concentration for various population groups (e.g., regions, ages, rural, etc.) should be estimated with an error rate less than y%.
- When the percentage of persons with concentrations of a particular substance above some threshold is compared at two times (e.g., this year and last year, or this year and 5 years ago), any change over y% should be recognized as signaling a real change over time and not a possible artifact of sampling error, and it should be measured with an error rate no greater than x%.

Other examples could be given, but these questions may the most critical kinds of data analysis. It is necessary for EPA to consider such issues before statistical theory can be applied to calculate the sample sizes appropriate for the NHATS.

The questions above focus on describing realistic goals for the study. They are intended to permit calculations of sample sizes that are sufficient for the analytic needs of NHATS and simultaneously to avoid using sample sizes that are larger than required for sensible analysis. For example, in comparing changes over time, it is necessary only to detect differences that are of practical importance—not all differences, but practical importance has to be defined.

In the context of the changes over time in NHATS data, that means deciding the changes in contamination that would be large enough to take to the attention of EPA policy-makers and others concerned with environmental quality.

Once the questions have been formulated and answered, it is fairly straightforward to calculate the necessary sample sizes. For example, in regard to the first two questions, an approximation to the standard error, σ, of a proportion can be stated this way:

$$\sigma = \sqrt{(dpq)/n}$$

where

σ = standard error desired (in terms of the total population, i.e., $p \pm \sigma$; not in terms of the relative error of p);

p = an estimate of the proportion of persons with detectable contaminant, or of the proportion with concentrations above some threshold;

q = $1 - p$;

n = sample size; and

d = design effect arising from complexities in the sample design.

The value of σ should be based on a realistic assessment of the precision necessary for the study. p can be estimated from past data or other previous studies. The design effects will not be known in advance of introducing a probability sample. On the basis of evidence from other health studies, we think that $d = 2$ would be a reasonable, although rough, approximation. The sample size can then be calculated by inverting the above formula to

$$n = dpq/\sigma^2$$

The formula is simple to apply. The formula for calculating sample sizes when two periods are compared, however, is somewhat more complicated, but tables are available (Fleiss, 1981). A more general formula can be used for mean concentration rates:

$$\sigma(\frac{2}{x}) - d(s^2/n),$$

where

$\sigma(\frac{2}{x})$ is the standard error desired, and

s is the standard deviation of concentration rates of the population.

We present two examples of how the formulas and tables can be used to determine the sample sizes necessary for expected analyses of the data. (These examples are meant to be illustrative and might not reflect actual NHATS requirements.)

• Analysts might believe that about 10% of the total U.S. population have detectable concentrations of a particular class of substances, although the exact amount is not known. They want to make sure with a high level of confidence that, if the proportion is really 10%, the survey result will be within two percentage points of 10%. The high level of confidence is defined as having a 95% chance that the survey is within 2% of the true value. The 95% chance implies that two standard errors is equal to 0.02, or $\sigma = 0.01$. The formula $n = dpq/\sigma^2$ with $d = 2$, results in a sample size of 1800. The same results can be obtained by reference to table 1. If one examines the column for $p = .1$ to find the value of .01, it appears on the line for $n = 900$. The table assumes that $d = 1$, so 900 needs to be multiplied by 2.

• Analysts might want to be sure that they can detect any major trends in the proportion of persons with detectable concentrations of some substance. They plan to carry out a simple test to see whether a change is occurring: comparing the average value of the proportion in the last 2 years with the average in the preceding 2 years. In the earlier period, the proportion was 20%. They want an 80% level of confidence that, if the proportion has gone up to 25% or down to 15%, the survey will reveal that a statistically significant change has occurred. Furthermore, they want to make sure, at the 95% level of confidence, that, if there has not been a change, the survey will not mislead them into believing that a statistically significant change has occurred. Statisticians refer to the 80% level of confidence as the power of the test. The 95% is denoted by 1 - α, so α = 0.05. Table A-3 (Fleiss, 1981) indicates that, to distinguish between values of p = 0.20 and 0.25 with σ = 0.05 and a power of 80%, a sample size of 1,134 for each of the two periods is necessary. The table (like most published tables of the type) assumes that d = 1. When d = 2, the necessary sample size is 2,268. The same table shows that, to distinguish between values of p of 0.20 and 0.15, a sample size of 945 is needed. Again, that has to multiplied by 2, so the sample size is 1,890. Obviously, the first condition is more stringent, and 2,268 are needed to satisfy both requirements. Because that is the sample size for a combined 2-year period, a sample of 1,134 per year is necessary. If the study is to have the specified power to detect *either* an increase to 25% *or* a decrease to 15%, one must use the larger sample size, 2,268.

Standard errors and sample sizes in the above formulas do not depend on population sizes. Although more exact formulas for estimating sample sizes do involve populations sizes (the ones above are close approximations), their effects are very small when the sample is a small part of the population, such as that of the United States, regions of the country, or large metropolitan areas. The practical effect of ignoring population size is trivial. That implies that, if one wishes to achieve a particular level of precision, the necessary sample size is independent of whether the population is that of the total United States, a region, or a subdomain defined in another way, such as males, rural residents, or persons at least 45 years old. As a result, the total sample size depends on the number of subdomains for which prespecified precision is needed (as well as the standard error desired). For example, as indicated earlier, for a standard error of 0.01 on a proportion that is expected to be 0.10, a sample size of 1,800 is necessary (with d = 2). If one wishes that precision in each of the four census regions, 1,800 are needed in each region, or 7,200 in the total United States. If the same precision is needed separately for males and females in each region, the sample size doubles to 14,400. Obviously, the sample sizes can go up rapidly with requirements for detailed analysis of subgroups, and data needs must be established carefully to keep study size and costs under control.

In the examples just given, the total sample size goes from 1,800 to 7,200 to 14,400. The sampling variances of national data will be reduced accordingly, so that when regional data are needed the U.S. variance will be one-fourth as large as when only national figures are required; when regional data by sex are needed, the U.S. sampling variances will be one-eighth as large. This direct arithmetic relationship holds only when the subdomains are approximately equal. When the subdomains vary greatly in size, there will be a much smaller reduction in the sampling variances of the U.S. totals. For example, if one wanted the same precision used earlier (a standard error of 0.01 on an estimated proportion of 0.10) separately for blacks and nonblacks, one would need 1,800 blacks and 1,800 nonblacks in the sample. However, the sampling variance on the U.S. statistics would not be reduced by a factor of 2. Because blacks constitute 12% of the U.S. population, having their sample size as great as for the nonblack 88% of the population does not provide as efficient a sample for the total population as having a sample of 3,600 distributed in accordance with the proportions in the population. In fact, the sampling variances on statistics for the total population will be reduced by only 21%.

Thus, sampling subdomains at different sample rates is better than sampling all subdomains at the same rate when one is interested in subgroups, but not as efficient for statistics on the total population. An approximation to the increase in variance arising from the variable sampling rates is Σ $(W_i k_i)$ Σ (W_i/k_i) where i is an index denoting the subdomains, W_i is the proportion of the population in the ith subdomain, and k_i is the sampling rate (or a factor proportionate to the sampling rate) in the ith subdomain (Kish, 1965). The increase in variance for totals should not necessarily deter one from oversampling to permit analyses of particular subgroups. For example, the current cycle of NHANES uses a wide variety of sampling rates for subdomains to be studied separately. However, decisions on oversampling should be reached only after careful, expert consideration of priorities and with recognition of the effect on costs and on national statistics.

Cooperation of Medical Examiners and Hospital Pathologists

We have not found data on the proportion of contacted medical examiners or hospital pathologists who refuse to cooperate or who drop out after initially agreeing, but the fact that no cooperators were available in 10 of the 47 MSAs indicates that the cooperation rate is quite low. In effect, volunteer medical examiners and pathologists, rather than EPA or its contractor, determine the sample. There is no way to know whether cadavers or patients in cooperating institutions are similar to those in noncooperating institutions. Most responsible survey organizations put considerable effort into attaining as high a response rate as feasible, and EPA, in its operation of the NHATS should have the same goal.

We believe that several steps could improve participation rates. First, the field personnel responsible for recruitment should be instructed on the importance of persuading reluctant prospects to cooperate. The NHATS should revise the manual currently used, which gives the impression that it does not make any difference who cooperates, as long as someone is found. Second, recruiters need tools to help to persuade reluctant prospects. One of the participants in the workshop (Appendix B) suggested that support from local medical societies might improve response rates. Another possibility is to try to get local health officials to help. Other health surveys, such as NHANES, generally enlist the support of such peer groups. Similar attempts for NHATS might be helpful. Finally, letters or telephone calls from high-level EPA officials could be effective. Invited respondents should know that the highest administrators at EPA recognize the NHATS as an important program.

Adherence to Quotas

We see no reason why medical examiners and hospital pathologists should regularly fall so far short of their assigned and accepted quotas. The MRI Operations Manual instructs field personnel to make calls to remind cooperating pathologists to check on progress, but does not emphasize the importance of meeting the quotas. Perhaps the field staff itself does not recognize that this is a major requirement or is too diffident in its dealings with medical examiners and pathologists.

EPA and MRI should examine their procedures to see what might be done to attain the desired quotas.

Replacement of MSAs

If the procedures described above increase response rates substantially, it would be useful to go back to pathologists in the discarded MSAs to see whether they can be persuaded to reverse their original refusal to cooperate. Those who agree should be included in the sample, and the replacements dropped.

In addition, EPA should evaluate the impact of the replacements on the statistics. One way is to examine differences in average contamination concentrations among SMSAs that are selected from similar strata (e.g., drawn from neighboring geographic divisions). Small differences would support the hypothesis that the SMSAs are fairly homogeneous within broad geographic areas; replacements then might not seriously affect the statistics. Larger differences could indicate a major effect of replacements on statistics.

Such comparisons can best be made by calculating components of variance, i.e., estimating what part of the total sampling variance comes from sampling MSAs and what part from sampling hospitals and persons within the SMSAs. To estimate variance components, one must measure the variability of contaminants among persons within the same MSAs. That cannot be done when composite samples are used, because the composites merge the data on individuals. It will be necessary to use historical data—from years when individual specimens were individually analyzed.

Nonrandom Selection of Counties, Medical Examiners, and Hospitals

In data collection, one does not need an uncontrolled system of choosing counties and institutions within counties. We believe that acceptable sampling methods could have been developed for those stages of sample. The current method turns over the selection to the subjective choices of field personnel. Reasons for that approach should be reviewed. If it was for some minor conveniences of the contractor and field personnel, plans for the basic samples within the SMSAs should be revised. It would also be useful to evaluate possible effects of the lack of specificity in the sampling. Some light might be shed on the subject by examining differences in contamination concentrations among hospitals in the same SMSA, or by comparing data on SMSAs before and after some hospitals were changed.

Composite Specimens

Starting with tissues collected in 1982, the EPA stopped analyzing adipose tissue specimens from separate individuals and started analyzing "composites." Individual specimens for immediate analysis (but not archived specimens) were combined within each of the nine U.S. Census divisions for three age groups. Some combinations were kept separate by sex or race, depending on the sample size in the census division. The combined tissues were mixed to form 46 "composites," which were the units subjected to chemical analysis. A linear model was fitted to the results of chemical analyses of the composite samples to derive estimates of average residue concentrations by geographic region, age, sex, and race in the broad scan work.

Two reasons were given for compositing:

1. To accumulate sufficient tissue in a sample to be analyzed to ensure a

high probability of detecting of toxic residues of interest. EPA indicated that that was necessary because the probability of detection is a function of the amount of analyte injected into the analytic instrument, in addition to the concentration.

2. To reduce costs of analysis by reducing the number of analyses. That was considered desirable because the cost per analysis is high. (For example, current dioxin analyses cost $1,500-2,000 per analysis.)

We discuss the amount of adipose tissue that can be extracted from a person in Chapter 5. EPA, through the NHATS, does not appear to be hindered in collecting sufficient tissue for chemical analyses of individuals. If amounts of tissue were not sufficient in the past, amounts can probably be increased in the future. Compositing will still help to reduce costs, however. Although individual specimens must still be selected and prepared, it obviously costs less to analyze a small number of composites than the much larger number of individual specimens.

The NHATS staff recognizes that costs are reduced at the expense of a substantial loss in ability to analyze and interpret NHATS results. Because individual specimens are not chemically analyzed, there is no direct way to derive prevalence estimates (e.g., the proportion of the population with detectable concentrations of some specific substance above a specified level). Statistical models from which prevalence can be computed have been discussed (Nisselson, 1987) but we are not sanguine about the prospects. However, if such models are attempted, their validity should be tested empirically. Perhaps it could be done by examining the data for the years in which individual specimens were analyzed separately.

Statistical modeling is currently used to estimate mean concentrations by geographic regions, three age groups, sex, and race. Direct estimates are not possible, because the specimens included in each composite cut across the subpopulations. The statistical model assumes that the mean log concentration for any combination of the four variables—region, age, race, and sex—is the simple sum of individual concentration factors for each variable and that there are no "interactions," or factors for combinations of the variable. For example, a model without interaction implies that the difference in the concentration of residues between males and females (although not the concentrations themselves) is the same for Caucasians and non-Caucasians and that the difference between Caucasians and non-Caucasians is the same in males and females. Similarly, lack of interaction implies that the sex difference and the race difference are the same in all regions. Nisselson (1987) has described the models and the mathematical expressions for the effects of interaction, but the adequacy of the models has not been tested sufficiently. As pointed out by Nisselson, the presence of interactions can drastically affect both subpopulation estimates and their standard errors. For the model evaluation, we suggest applying the model to a period in which individual specimens were analyzed and comparing the model results with estimates made from the individual observations.

COMPUTATION OF SAMPLING ERRORS

There has been little or no effort to establish statistical confidence levels around the estimates that have been prepared from NHATS data. It is therefore uncertain whether differences observed among subpopulations or changes over time are more likely to reflect real changes in exposure than random fluctuations among the sample specimens. Similar uncertainties arise when NHATS data are compared with data from other populations.

When essentially nonprobability samples are used (as in the NHATS), confidence intervals computed by standard statistical techniques do not adequately reflect the probability that the true values are within specified neighborhoods of the sample estimates. However, they are the best approximations that are possible and generally provide lower bounds on uncertainty. We suggest that the necessary computations be carried out for at least some of the substances analyzed and that the resulting information be made available both to the users of the data and to EPA personnel for consideration of the adequacy of NHATS sample sizes. Techniques can be used to estimate standard errors (EPA, 1987a). The computation of standard errors in most statistical software packages is not appropriate for a multistage sample design that uses variable weights.

CRITICISMS OF NHATS STATISTICAL SAMPLING METHODS

The EPA staff involved in the planning of the NHATS are well aware of the limitations of and problems with the NHATS sample. However, they believe that the NHATS nevertheless meets EPA's goals for the National Human Monitoring Program (see Chapter 2) as stated in their preliminary response to NRC inquiries; we would agree if those goals are considered in limited manner. The dramatic changes that the NHATS showed in the prevalence of PCB's in human tissue over a few years seem to reflect real changes in our environment. We would also generally assume that NHATS statistics showing increases of 100% or more in mean concentrations of some substances over a few years indicate important changes in our environment. Similarly, the NHATS can demonstrate even very low concentrations of some toxic substances in human tissue.

The problem with a mostly nonprobability sample is its inability to ascertain, with any measurable level of confidence, smaller annual changes in average concentration or population distributions that build up over time to have important effects in our society. With the present program, it cannot be determined whether moderate increases over several years indicate an increasing prevalence of higher body burden within the overall population sampled, a greater concentration in a small fraction of the population, erratic sampling effects, or a shift in the magnitude of the sampling bias due to lack of control on the sample. It could be a long time before the statistics show clearly recognizable patterns.

There are other problems in extrapolating from the NHATS sample to the total U.S. population. As mentioned earlier, one result of using composite samples is the loss of prevalence estimates. Composites might also create poorer estimates for subpopulations because of a need to rely on statistical models if compositing across subpopulations is used. The use of baseline data for comparison with results of studies of other population groups (e.g., persons living near Superfund sites, or those living in rural areas and subject to substantial pesticide exposure) is also weakened by the wide margins of uncertainty around the baseline statistics.

The current operating procedures probably do meet some set of limited EPA objectives. Those objectives would be met better if the improvements suggested above were adopted. However, we do not think that EPA, Congress, or the public should be satisfied with such limited objectives. The federal government should assume a more comprehensive responsibility for informing the public and administrative agencies about potential public-health hazards as revealed by the accumulation of chemicals in the population. That will require fundamental changes in data collection. We discuss some major changes below.

The Benefits of Blood Collection
for Probability Sampling

From a statistical point of view, a troublesome aspect of the entire NHATS program has been replacing the U.S. population as the target population with surgical patients and cadavers subject to autopsy. The replacement creates three serious problems: specimens taken in this way might well not represent the average living population, it is very difficult (probably impossible) to ensure a true probability sample, and it is not feasible to obtain important demographic or other data on the subjects.

Those problems are inevitable with a measurement system that uses human adipose tissue. Therefore, we have recommended that blood replace adipose tissue as the primary medium for measuring toxic, or potentially toxic, substances in human tissue. Blood specimens could be collected, in accord with procedures roughly similar to those in NHANES, from subjects that are close to a true probability sample of the U.S. population. We recognize that blood and adipose tissue differ in buildup of substances. However, the ability to have a good sample that would be accompanied by demographic and related descriptive material about each specimen makes blood the tissue of choice. Chapter 3 discusses the relationship of substances in blood to those in adipose tissues.

The use of blood specimens from a sample of live persons has other advantages. First, it would be possible to conduct interviews with the sampled persons to obtain information on covariates that would support a search for causal relationships and risk factors. The covariates could include type of drinking water used by the household (private wellwater use vs. a community system), occupation and industry (particularly whether employment is in a chemical plant or refinery), and farm vs. nonfarm residence (and, if on a farm, use of pesticides, dietary information, etc.).

Second is the possibility of a longitudinal design in which sampled persons (all or some) are revisited every few years, instead of being selected independently each year. Such a design has two desirable features. It usually provides more precise estimates of year-to-year changes. And, it permits more sophisticated analysis of sources of contamination by attempting to associate changes in the concentrations of toxic substances in a person with changes in the environment (e.g., construction of a new road or industrial facility) or in other factors peculiar to the person (e.g., a job or residence change). We note, however, that a longitudinal survey also has some disadvantages. It is usually expensive to locate and visit the part of the sample that has moved between sample periods. Some movers cannot be located or have moved to areas that would be inordinately expensive to visit. Response rates thus tend to decrease over successive rounds of followup. Finally, there is a loss of ability to increase sample size for some analyses by combining data for several years. Thus, we do not unconditionally recommend introducing longitudinal features in the sample. However, the advantages and disadvantages should be carefully weighed, so that a reasonable decision on the best sample design is reached.

Third is the possibility of considerable flexibility in oversampling specific demographic or other subgroups of the population. There is some oversampling in the current NHATS, in that quotas are specified by sex, race, and three age groups. That could be extended to a finer division of age groups or to other domains. Some possibilities are the rural population, persons living in areas with heavy pesticide use, and persons living in the vicinity of particular types of industrial facilities.

Despite the value of those additional features of a program based on blood, it must be clearly understood that the primary rationale is the need for true probability sampling, and that the advantages of having a probability sample must be protected in other aspects of study design. For example, EPA's plans for a National Blood Network (NBN) are based on the collection of blood specimens, through the cooperation of the major national blood collection agencies, from volunteer donors to those agencies. Such a program would have many of the problems inherent in the NHATS, because it would rest on the assumption that the population of volunteer blood donors is a useful surrogate for the general U.S. population—an unvalidated and even dubious assumption.

Furthermore, organizations whose priorities are in their own programs generally do not give other projects close attention. It is likely that the NBN will be subject to many of the NHATS operating problems, such as the refusal of sampled units to cooperate, inability to meet quotas, and general lack of quality control. Those problems are not conducive to the high quality statistical program that the public has a right to expect from EPA. EPA should plan to have a sample and data collection system that is dedicated to EPA's interests. NHANES provides a good model. We recognize that the costs to EPA would be much higher than the current budget; however, we believe that EPA has understated the need and importance of the data and thereby has been too modest in its budget requests and allocations.

In addition to a blood program, we recommend that, in spite of the statistical limitations of the NHATS program, an adipose tissue program be continued, although possibly on a reduced scale. One reason is to retain the consistent series of specimens that goes back almost 20 years. A second reason is that the NHATS would supplement the blood analysis program and provide data on some substances that cannot be measured adequately in blood. If an NHATS program is retained, the improvements and modifications described elsewhere in this document should be implemented.

Use of Composite Specimens

The use of composites has required EPA to set aside some of the objectives initially stated for the NHATS. Prevalence estimates can no longer be supplied. Estimates of mean concentrations for subpopulations are less precise, because they assume model validity. The ability to carry out risk assessment is uncertain. Those compromises in the initially contemplated program have been made to permit broad scan analysis of an increased number of substances, which becomes quite expensive per specimen. Although the broad scan analysis provides a substantial amount of information not available with the original measurement method, it does not satisfy all the purposes of a human tissue monitoring program. In deciding to use broad scan analysis, EPA acted as though it had a fixed budget for monitoring and as though the higher costs per chemical analysis had to be compensated for with a smaller number of samples analyzed.

The realities of life in a government agency are such that current programs need to work within fixed budgets and it is difficult to change them, except under extraordinary conditions. However, those conditions do not necessarily apply to long-range programs. It is possible to request and obtain increased funding when it is necessary for the success of an important project. In fact, government agencies are obliged to work to obtain such increases.

EPA should consider the benefits of a mix of analyses—some performed on individual specimens and others on composites—in which the results from individual specimens can be used for prevalence estimates and to test the models. It might not be necessary to prepare prevalence estimates each year; if not, smaller samples can be used for the individual specimens, with prevalence estimates based on 2- or 3-year averages. The precision required for the most important uses of the data should be reconsidered, and the total sample sizes and the allocations to samples used for individual specimens and for composites should be recalculated to meet these requirements.

Samples should be collected and stored in a manner that preserves the possibility of basing measurements on individual samples, and a substantial part of the new program should be based on individual analyses. Compositing can reduce costs at the stage of chemical analysis and thus permit additional sampling or studies. When it can be shown explicitly that values based on individual samples are not needed (e.g., for estimating variances, presence above some specified concentration, or differences among population segments), some degree of compositing might be appropriate, though that degree may never reach the present EPA degree of compositing.

NHATS Sample Design

If our primary recommendation for a new program is not adopted, and if the present program is continued with modifications, its sample design should be reviewed and revised to institute probability sampling at all stages of sampling. Key aspects of the review should be as follows.

• Nonmetropolitan counties should be included in the sample. The rural and urban populations may well differ in prevalence and body concentration of pesticides and possibly of other substances. We recognize that hospitals in rural counties are generally smaller and that medical examiners have fewer cases, but populations are smaller, too. Thus, quotas for the number of specimens in those rural counties may be smaller than in SMSAs. That will add some fixed costs per hospital to the project, but is necessary for true representation of the U.S. population.

• The quotas for the sample size for each SMSA should be updated after the 1990 Census results become available. Current quotas are based on the population distribution in 1980, and there have been important changes in the last decade.

• The subjective method of choosing counties (within the selected SMSAs), medical examiners, and hospitals should be replaced with strict probability sampling.

• Stronger efforts should be made to attain cooperation of sample institutions and to have them meet their sample-size goals.

• It might be difficult to have the medical examiners and pathologists select specimens with random sampling methods, but the possibility should be explored, particularly in the larger organizations. As a minimum, an attempt should be made to spread the selection of cases evenly across the year, to avoid possible seasonal effects.

A broad national program for human tissue monitoring might at first seem to be a suitable vehicle for studies and evaluation of groups in the population that appear to have unusually high exposure to some chemical substances. Examples include accidental exposures to environmental disasters, occupational risks, and persons who live near Superfund sites. The committee believes that, in general, evaluation of other than background exposures should not be built into a national human monitoring program, although data collected in the broad basic program that turn out to be useful should of course be used. The basic problem is that a broad sample would include only a handful of persons in any small group of special concern, and design of a sample to answer questions about such groups would either require an enormous equal-probability sample or distort a weighted sample to the point where it might be unsuitable for its basic purpose, even with appropriate analyses to adjust for differential weights.

An agency staff qualified to maintain a strong human tissue monitoring program should have skills and facilities for investigation of risks in special populations, and special studies might well be assigned to them. However, such special studies should in general be regarded as add-ons, supported by separate budgets and explicit reallocations of staff responsibilities and time commitments, so that the basic monitoring program is protected. The national data will, of course, be invaluable for comparison with results of special studies and might determine whether the special populations have increased concentrations of contaminants.

SUMMARY AND RECOMMENDATIONS

There are serious deficiencies in the NHATS in that adipose tissues collected are not a representative sample of the U.S. population. Although most population surveys find it necessary to compromise somewhat on ideal standards, the departures from probability sampling in the NHATS are far in excess of what most statisticians would consider acceptable. The main deficiencies are these:

Although the target population is the living U.S. population, the subjects on whom measurements are taken are an uncontrolled mix of recently deceased persons and surgical patients.

• The sample size has been driven by the budget, rather than by needs to satisfy important goals of the program.

• Some important segments of the population are omitted from the sample. The exclusion of the rural population is the most serious omission.

• Although probability sampling was used in the selection of the metropolitan areas that are the first stage of sampling, problems of cooperation forced substitutions for 20% of the areas. Consequently, the extent to which the sample of areas now represents all metropolitan areas is uncertain.

• There is no designated sampling method for choosing the persons from whom specimens are taken. Each medical examiner or pathologist is given a quota by age, sex, and race, but the quotas are poorly adhered to; even if the quotas were met, the quota procedure would be inherently biased.

• There are no specific instructions for pathologists on the body part to be used for specimens. It is implicitly assumed that contaminant concentrations are the same in all adipose tissue in the body.

• Recent uses of composite measurements have made it impossible to provide prevalence estimates and seriously weakened the estimates of mean contamination concentrations for the sex, race, and age subdomains.

• Sampling errors have not been calculated, so users are not informed about the precision of the data.

• There is no plan for regular release of findings to the public.

Even within the limitations of the NHATS system, it is possible to improve the methods used to implement the NHATS so that it can come closer to a realization of the original plan. This chapter includes recommendations for improvements. Even with the improvements, however, the NHATS will have major limitations in its ability to reflect the accumulation of toxic substances in the U.S. population. The committee believes that the limitations are so serious that the NHATS should be replaced with another system for measuring the accumulation of contamination in human bodies.

The committee recommends that the NHATS be replaced with a blood monitoring program as the primary method of measuring toxic or potentially toxic substances in human tissue. The sampling plan can be patterned after the one used in NHANES, whose subjects are close to a true probability sample of the U.S. population.

In addition to having a probability sample, the system would permit interviews to be conducted with the sampled persons to obtain data on covariates. Studies could then be made on sources of contamination as well as on amounts in human tissue. Other advantages of using blood are described in Chapter 3.

Collection of blood specimens should be designed in strict accord with the methods of probability sampling at all stages. The methods used should be efficient for giving virtually all persons in the United States a known probability of selection.

During the period in which the NHATS is continued, the selection methods should be revised to reduce the subjective elements in the choices of counties, hospitals, and specimens.

In addition to initiating a blood collection program, the committee recommends the continuation of the collection and analysis of adipose tissue, although possibly on a reduced scale.

One reason is to have a continuous time series, and a second is to provide data on substances that cannot be measured adequately in blood.

5

Collection, Short-Term Storage, and Archiving of Tissues

INTRODUCTION

The committee evaluated EPA contractors' current methods of collecting and storing human tissues and visited the facility that houses the NHATS samples to examine those tissue specimens. This chapter addresses the issues to be considered in the collection, short-term storage, and archiving (long-term storage) of tissues for chemical analysis. It summarizes the committee's findings and presents its recommendations.

COLLECTION

Many programs collect tissues for monitoring a population's exposure to environmental chemicals and other xenobiotic substances; the design and details of specific programs have been discussed in various publications (Wise and Zeisler, 1984; Lewis et al,. 1987). Material for monitoring may be obtained by relatively noninvasive techniques—collection of blood, urine, hair, fingernail clippings, etc. The National Health and Nutrition Examination Survey design provides a model for the collection of human tissues and is described in Appendix D.

A program that monitors environmental chemicals through analysis of chemical concentrations in solid tissues, such as fat or liver, requires a tissue collection network. The design of a network depends on the overall goals of the program. Establishing a collection network is a complex process, and recruitment of field personnel committed to the goals of the program is critical to its success. Close cooperation and communication between program managers, contractors, and field personnel (i.e., primary collectors) are necessary; a successful collection network is likely to require decentralized responsibilities for operating units. For instance, a primary collector might be responsible for collecting specimens from all facilities in a demographic (e.g., pediatric) or geographic (e.g., rural) area. A contractor responsible for a multistate area might be more successful in recruitment and training of collectors within that area than a contractor that tries to control collections nationwide from a single location. The contractors' and program managers' relationships with collectors should not be taken lightly; once potential collectors are identified, long-term relationships should be established among collectors, primary suppliers, contractors, and program managers, because intermittent collection does not encourage continued participation by collectors and primary suppliers.

In a national human monitoring program, the selection of institutions to participate in the collection of tissues depends on both the characteristics of the persons from whom tissues are to be collected and the types of tissues needed. For example, if tissues are obtained primarily from accident victims who die very soon after their injury, collection is often restricted to coroners and medical examiners in designated geographic locations. But medical examiners might autopsy patients without family permission; in some states, such tissues might not be available for collection. Additionally, autopsies might not include examination of internal organs, if the cause of death is obvious from external examination. Finally, many traumatic or accidental deaths involve drug abuse, and use of samples from such deaths could compromise the representativeness of a tissue collection. For those reasons, it is unwise to restrict tissue donors to persons who die traumatically.

If tissues are collected from a wider spectrum of patients, it is important to specify eligibility requirements with respect to length of illness, nutritional state, active disease processes, and whether surgical or autopsy specimens are to be collected. Researchers using data from collection network must know the range of pathophysiologic states that affect collected tissues. Too broad a spectrum of eligible issue donors could lead to the collection of misleading or unusable tissue specimens; too narrow a spectrum could produce an unrepresentative collection of specimens.

A well-coordinated collection network cannot operate optimally with telephone communication as the primary method of contact. Frequent visits by program managers and contractors to collectors are necessary, in addition to periodic training sessions for all personnel involved in collecting tissues. Training sessions should emphasize proper specimen collection (e.g., use of storage containers that will protect specimens during shipping and storage and use of instruments that will not contaminate specimens), protection of personnel from contamination by infective agents, and quality control. During training sessions, collectors should be informed about the quality of their previous samples. Instruction should be provided on obtaining medical and social histories of persons from whom tissue samples are taken. Those activities at training sessions should result in more consistent sample preparation and sample data and in better quality control of specimens.

Optimal collection of histories of specimens from living donors requires a standardized medical-history form (including the reason for surgery in the case of a specimen taken during surgery); additional information is required for autopsy specimens (e.g., the final anatomic diagnosis and the lengths of the terminal illness and final hospitalization). Even if not all information is available on a given specimen, the sample might still be usable; if samples were excluded for lack of data, the representativeness of the collection might be compromised and specimen availability limited.

SHORT- AND LONG-TERM STORAGE

Proper conditions are necessary for short- and long-term storage of tissue specimens. For instance, the integrity of tissues must be maintained in storage to enable valid chemical analyses. Storage freezers that keep temperatures above -80°C and that have a defrost cycle might compromise specimens and chemical analyses (see discussion below), and freezing and thawing of solid tissue specimens can result in enzyme release into cells and destruction of cell membranes and cell integrity.

Quick freezing of specimens is essential. If collectors or collection facilities do not have the capability to freeze specimens rapidly and ship them to contractors (for chemical analyses or archiving) on the day of collection, the tissue monitoring program or its contractors might have to supply collectors with containers and freezers for proper storage. Supplying each collector with a small -80°C freezer would result in a one-time cost of about $5,000 per freezer. The freezers could be owned and maintained by the monitoring program or its contractors, and they could be transferred to new collectors or facilities as necessary. That approach would permit temporary storage of specimens without major degradation. Unfortunately, freezing and thawing of solid tissues can destroy cell membranes and release enzymes into tissues, even if appropriate techniques and equipment are used, and that can adversely affect chemical analyses and identification of biologic markers that could potentially provide information on an environmental exposure.

Archiving or banking of specimens and tissues consists of the systematic collection and long-term storage of selected organisms or tissues. Many authorities in environmental monitoring believe that a prospectively designed tissue bank should be part of environmental monitoring programs (Lewis et al., 1987), although retrospective analysis will also be an important function for an archive. The advantages of archiving tissues in conjunction with monitoring programs have been summarized in several publications (Kayser, 1982; Martin and Coughtrey, 1982; Lewis et al., 1987) and include the following:

- It permits continual evaluation of the distribution and concentration of chemical residues.
- Chemicals that have had little attention in the past might suddenly warrant investigation. Archived samples would be required to identify times of first appearance, sources, and trends of such chemicals in the population.
- As methods advance, results of prior assays might no longer be considered adequate to detect or discriminate among specific compounds or to provide adequate quantitation.
- It is more economical and scientifically sound to maintain a banking system than to perform assays that are extremely broad. Even with the most comprehensive assay methods, many compounds that will be important in the future might not be identified, let alone measured, and some that might be unimportant in the future would be analyzed in more depth than is needed.
- Banked portions of previously analyzed specimens can aid in the development and quality control of analytic methods.
- Banked specimens can be analyzed in parallel with current specimens to provide accurate assessments of the efficacy of regulatory actions.

Major banks of human tissues can be found in the Federal Republic of Germany, where blood, liver, and adipose tissue are banked (Kayser, 1982; Lewis et al., 1987), and in the United States, primarily in NHATS (adipose tissue) and the National Institute of Standards and Technology's program (liver) (Wise and Zeisler, 1984; Wise et al., 1988). Studies of human-specimen banks are under way in Sweden (Andersson and Gustafsson, 1989) and Japan (Ambe, 1984). In the United States, numerous specimen banks for environmental monitoring have been developed to collect nonhuman specimens, including herring gull eggs (Elliott, 1984), bald eagle tissues (Stafford et al., 1978), fish (Schmitt et al., 1983), shellfish, including mussels (Wise and Zeisler, 1984; NOAA, 1989) and oysters (Wise et al., 1988), and sediments (Wise and Zeisler, 1984; NOAA, 1988). The Canadian government conducts an extensive monitoring program in which a wide variety of animal tissues are banked for environmental studies.

Quality Control

Each collector must provide for quality control of the specimens provided. One form of quality control is the preparation of a hematoxylin and eosin (H&E) slide from a portion of each specimen. That procedure enables a pathologist to determine the suitability of a tissue for analysis. For example, it would enable a pathologist to confirm cell and tissue type, to identify necrotic tissue, and to identify atypical cells in the specimen. Quality-control procedures should prevent specimens that are unsuitable (e.g., autolyzed or otherwise compromised during collection and storage) from being accepted by the program.

Contractors should also design an overall quality-control program for tissue collection and archiving. At a minimum, the quality-control program should provide for collector training, periodic monitoring of collectors, sample-rejection criteria (e.g., disease exclusion, and improper shipping), and monitoring of archiving activities.

Specimen Size

Although the committee cannot recommend a specific sample size, the size of a tissue specimen to be collected should be based on several considerations. One is the type of tissue; for example, assays of adipose tissue often require small quantities (5-20 gm) where as assays of blood might require 1-200 ml per assay (elemental screening and dioxin analyses are extreme examples). Others are the multiple uses of the tissues envisioned for the program, the quantity of tissue needed for chemical assay, the constraints of collection and storage mechanisms, and cost. Additional uses of the samples might be warranted and could include archiving for future retrospective analyses and development and for validation of analytic methods and use in special studies by other agencies or researchers; choice of sample size should account for such possibilities.

ADDITIONAL ISSUES FOR CONSIDERATION

The goals of a human-tissue bank, operated in parallel with an environmental monitoring program, should be clear and should be supported—with a long-term commitment—by the organization sponsoring the program. The design of the archive should enable the goals to be realized and should enable whatever flexibility is warranted by the program. A specimen bank organized in conjunction with environmental monitoring will have many purposes, but should support the following: evaluation or formulation of regulatory policies concerning chemicals that might affect the environment, making of decisions about modifying local and regional environmental burdens of selected chemicals, illumination of issues in environmental litigation, risk-benefit analysis of environmental regulations and legislation, and direction in environmental research. The design of the tissue bank should enable the program to address those issues, but its utility will depend on its correct operation, its resources, and the proper selection, collection, handling, and storage of specimens.

Specimens for Tissue Banking

Considerations regarding the type of specimens to collect for an archive are the same as those regarding specimens for environmental monitoring and depend on the objectives of the program. The tissues banked should be a subset of the tissues collected for the concomitant monitoring program. The types of human tissues to be collected for environmental monitoring have been discussed in other documents (Lewis et al., 1987). As discussed in Chapter 3, different tissues act as reservoirs for different chemicals and many retrospective analyses involve measurement of chemicals not yet designated, so it is important to bank more than one tissue type. For example, if only adipose tissues were stored and a class of nonlipophilic compounds were identified requiring analysis, valid measurements would be impossible, because the compounds would not concentrate in lipid.

The environmental-specimen banking program of the Federal Republic of Germany has elected to analyze and bank three types of human tissue—blood, liver, and adipose tissue—and nonhuman specimens to enable chemical analyses of a broad range of contaminants (Lewis et al., 1987). However, specification of "adipose" or "liver" tissue might not be sufficient for uniformity of analysis, because a given kind of tissue from different sites can vary in ability to store various chemicals or chemical metabolites. Thus, specific locations, such as "perirenal fat" or "periumbilical fat" might need to be stipulated in collection protocols. Location also needs to be considered in the design of programs.

Tissue Storage Conditions

Temperature

If chemicals of many classes are to be measured accurately, storage conditions (even in short-term storage) must be such as to maintain cell integrity and prevent degradation. Most tissue banks store specimens at -80°C or lower (Wise and Zeisler, 1984; Lewis et al., 1987; NOAA, 1988); the German bank is maintained at -171°C or lower. Ideally, tissue banks should store specimens at the temperature of the vapor phase of liquid nitrogen (-150°C) because enzymes that can be active at -80°C will not be active at the other very low temperature in common use, -150°C, and some rely on chemicals and biologic markers that are stable only at -150°C and lower (Nürnberg, 1984). Thus, environmental monitoring and banking programs should elect liquid-nitrogen storage temperature. Low-temperature storage (-80°C or lower) should be used only for temporary storage at collecting sites. Specimens should never be stored in self-defrosting freezers, which accelerate desiccation ("freezer burn"). Properly frozen specimens can be shipped to a central storage facility on dry ice, if appropriate procedures are followed to ensure proper temperature and specimen freezing are maintained.

Storage Containers and Specimen Size

Storage containers should be chosen to prevent contamination of specimens, be structurally stable, and ensure specimen stability at liquid-nitrogen temperatures. The National Institute of Standards and Technology elected to use Teflon containers for the storage of liver specimens in liquid nitrogen, whereas the special characteristics of blood required that it be collected in one type of container and stored in another (Wise and Zeisler, 1984; Wise and Zeisler, 1985). When whole blood is banked, an anticoagulant might be needed, not only to prevent coagulation, but to enable separation of plasma and red cells, which may then be stored separately (freezing of whole blood leads to lysing of red cells). Freezing and thawing of samples should be kept to a minimum. A freeze-thaw cycle might change chemical partitions between phases in tissues, introduce contaminants, or damage biologic markers, and thus make later analysis unreliable. That can be avoided by dividing specimens into aliquots and freezing them in separate containers, perhaps as homogenates prepared cryogenically (Wise and Zeisler, 1985). Aliquots of specimens could then be thawed as necessary for chemical analysis.

Analytic error is reduced if specimens are 1 g or larger (Wise and Zeisler, 1985); aliquots of 1-2 g might be maybe optimal. However, it is important to store large specimens in small aliquots and containers to minimize uneven thawing. For example, a 15-g specimen of adipose or other tissue would be 3-4 cm in diameter, and the outside of such a specimen would thaw well before the center. That could be prevented by storing the specimen in aliquots that would thaw more uniformly.

Although we cannot recommend an optimal size for specimens, the factors to consider include the minimal size needed for detection of important concentrations of a particular chemical, foreseeable demand for the specimen, ability to collect samples of specified sizes, storage space, and storage resources. "Minimal size" cannot be defined, because an archive will eventually provide specimens for analysis of chemicals not specified at the time of collection. Specimens should be as large as is consonant with problems of collection, storage space, and resources. A "reasonable" size of specimen of fat or liver to request from a medical examiner performing an autopsy is 15-20 g; requests for whole-blood samples from living persons could be at least 15-20 ml.

Archival Information

Archival records are important, not only for tracking and locating individual specimens, but also for following the storage histories of individual specimens and for making decisions on specimen use. For example, if a specimen was divided into 10 aliquots, and nine have been used, a decision to use the remaining aliquot requires careful consideration. The need to archive specimens and their potential multiple uses require that samples be stored in two or more containers. A collection network must establish clear, unambiguous methods for identification of samples (including aliquots) that will be used consistently throughout the network and that will protect donor confidentiality.

Access to Archived Specimens

Access to specimens in an archive must be carefully controlled. In each case, it must be determined whether a projected use will provide useful data and its value must be balanced against the need to maintain specimens for future studies. An independent scientific group can review requests to use archived specimens. The availability of archived specimens for extramural projects, as well as potential funding for those projects, should be advertised by EPA. EPA's Scientific Advisory Board or the independent oversight committee that we recommend elsewhere could assist in that role of selecting extramural projects.

STATUS OF THE NHATS ARCHIVE

History

The archive of the NHATS was initially composed of the remnant portions of specimens collected by the Centers for Disease Control (CDC) to evaluate pesticides in the environment. The tissue collection was not originally planned as an archive, but it became an archive because investigators did not want to dispose of remnant tissues that were considered a valuable resource. The archive is currently maintained by the Midwest Research Institute (MRI) under contract with EPA.

Current Status

NHATS specimens are stored in two locations. Early specimens (those collected through 1984) are stored in an underground warehouse maintained by Americold, an MRI subcontractor. They are maintained at -16 to -18°C. Many of the specimens were transferred to the EPA contractor from the CDC program, and their exact histories are unknown. However, knowledge of two episodes of power loss at a prior storage site suggests that some specimens have thawed at least twice. The specimens are in their initial specimen bottles, which are in Ziploc® bags that are stapled shut. They are grouped roughly according to year of collection. These bags are stored in large cardboard boxes (approximately 8 feet³ in volume), each of which is approximately half full of specimens from a 2- to 3-year period.

Committee members visited the warehouse to inspect and evaluate the condition of the archive on January 17, 1989. Representatives of EPA and MRI were present. Three boxes were opened and the contents inspected: one contained specimens from 1970-1972, one specimens from 1976-1979, and one specimens from 1981-1983.

Specimens from 1970-1972 had severe storage artifacts. The specimens were stored in glass bottles that had metal caps with foil-lined cardboard inserts. The foil lining of approximately 10-15% of the specimens had deteriorated extensively and the specimens contained pieces of foil. Some caps were rusted. In some cases, fluid or lipid from the specimen had leaked through the cap and stained the outside of the specimen container. In others, the cap was loose, and the specimens were dried out. Some specimens adhered to the tops of their containers, rather than the bottoms; that suggests past thawing of the specimen upside down, followed by freezing and adherence to the tops of containers. All samples were dull gray, instead of the expected yellow.

The manner in which many of the glass bottles were stored in the plastic bags (i.e., cap down) exacerbated storage problems--contact and probable contamination with the caps.

Specimens from 1976-1979 showed changes similar to those just described, but a smaller proportion of specimens contained metal foil and drying appeared less severe than in the 1970-1972 specimens. All specimens showed color changes from the usual yellow to dull gray and tan. The same general problems, including drying, existed with all specimens, and the changes were accentuated in small specimens and those whose container caps were loose.

The storage artifacts of specimens from 1981-1984 appeared to be still less extensive. However, some early foci of corrosion of the foil linings of caps were noted, and drying was apparent in smaller specimens and those with loose caps. Most specimens were grayish to yellowish tan, in contrast with the expected yellow.

The remaining specimens (i.e., those collected after 1983) are stored at MRI in self-defrosting freezers kept at about -20°C. Like those in the warehouse, many of the bottles are stored upside down in freezer bags, so specimens have made contact with caps.

We examined the specimens at MRI according to year of collection. Specimens collected most recently (1988) had the bright yellow color typical of adipose tissue. Two samples appeared to be contaminated with fungus or had developed a crystalline formation. Specimens from 1987 were no longer yellow, but were dull gray or tan like the older specimens in the warehouse.

In almost all cases, there was extensive ice-crystal formation both on the specimens and on the sides of the containers. The presence of ice crystals in the more recently collected specimens suggests freezing of moisture from within the containers that resulted from drying of the specimens. The extent of crystal formation varied with the geographic location of specimen collection.

Important questions are how to obtain "representative" aliquots of frozen samples for analysis and what quantity of a sample with crystalline formation to include in a "representative" aliquot. An important technical consideration is the effect of desiccation on the analysis of substances measured. Desiccation can affect concentrations of volatile chemical metabolites. Lipid-soluble chemicals might not escape with the moisture, but the fate of volatile substances, as well as substances with low partition fractions in lipid, is not as clear. Additional uncertainties in the analysis of specimens arise from contamination with foil (probably aluminum) from the specimen-bottle tops and from the lack of a clear history of the specimens before transfer to the current contractor.

Finally, although the specimens supposedly are primarily from persons who died accidentally, that might not always be the case. Some specimens might have come from persons subjected to toxic exposure, malnutrition, etc.

SUMMARY AND RECOMMENDATION

Artifacts of storage of NHATS specimens are seen within the first 6 months of sample receipt. Storage artifacts—including drying, fungal growth, and contamination with portions of the tops of specimen containers—increase with time and become severe after 5-6 years of storage; they affect almost all specimens after 10 years of storage. It is not clear that "representative" aliquots can be taken from the specimens. Histories of specimens and documentation of adequate storage conditions might be deficient and should be so recognized in publications that use data from the archive. A primary question is the extent to which these storage artifacts affect specific analyses and compromise the value of the NHATS archive. The effects of storage of specimens at temperatures above -20°C (self-defrosting) on future analyses are not well understood. Some of the effects might be reduced by the use of better sample containers, a better method of storage, more control of storage at the site of collection, and better documentation of the histories of specimens.

Given the state of the current archive, the committee believes that the existing frozen samples of adipose tissue are likely to have little or no value to a successor program or to other parties inside or outside government.

However, because the matter has not been adequately studied, the committee recommends that the archive be preserved until a successor program can give its use early consideration, specifically asking: Should the archive be saved indefinitely or discarded, and, if it is to be saved, how should it be preserved and how should it be used?

Newly collected specimens should be archived according to up-to-date protocols.

6

Chemical Assay of Specimens

INTRODUCTION

"Monitoring" implies routine measurement that is inherently closed-ended and based on established methods and practices. Monitoring programs vary with regard to requirements and approaches, but usually have as a base a list of analytes ("target chemicals") and assay methods that have been validated for the sample type and concentration range of interest. A successful monitoring program maintains results over time for comparison and must therefore be technically adequate at the outset. Comparability is most easily achieved if assay methods are constant. A monitoring program often offers a good setting for other kinds of activities, and the National Human Monitoring Program includes aspects that are not monitoring, such as recognition of new agents of concern and detection of chemicals not previously included in monitoring protocols. The latter objectives are appropriate to a population-based biologic surveillance program, but they require an approach to chemical analysis different from that for monitoring. Design of a program of surveillance requires a plan for development and change in analytic methods, as well as a plan for maintaining stable methods and analytic quality control. Successful balancing of the routine and the innovative or exploratory components of such a program is a major challenge for the program's management.

The following discussion of aspects of a program is based on several assumptions:

- Present knowledge does not permit designation of all substances that might be detectable in tissues or that would be important if detected.
- Present analytic technology is inadequate for surveillance of some sub-

stances, because of limitations in sensitivity, in applicability to some chemicals of interest, and in cost.

• Individual target chemicals will increase or decrease in importance over time, but relatively slowly. The slowness of increases in importance reflects the slowness of increases in US population-wide exposures. Decreases in importance reflect similarly slow decreases in exposures and increases in understanding of the nature of the health implications of and reasons for some exposures. "Emergency" exposure assessments that might be required when a serious health hazard is discovered would be best addressed in focused special studies.

• Monitoring efforts will provide opportunities for exploration of tissue composition beyond a list of target chemicals. Those opportunities and other efforts parallel to the monitoring tasks should support continuing development of the monitoring program itself.

MAJOR FEATURES OF A PROGRAM
FOR CHEMICAL ANALYSIS OF TISSUES
FOR POPULATION-BASED SURVEILLANCE OF EXPOSURES

Monitoring-Program Development

One must first establish goals, specify target chemicals and quantities, and identify analytic methods. The usual reason for establishing a monitoring program is recognition of a problem or potential problem related to known agents. Programs of environmental monitoring have used several kinds of information in formulating lists of target chemicals, such as case reports of environmental concentrations or industrial releases, volumes of chemicals produced or sold, toxicity or other hazard-ranking factors, and analytic feasibility. A chemical would be a good candidate for inclusion in a tissue monitoring program if it met the following requirements:

• The chemical is detectable with currently used or available methods.

• It would appear in the tissue at detectable concentrations in the event of exposure.

• Its presence in the tissue would indicate an exposure, i.e., would represent a marker of exposure or the actual substance or its metabolite would be present only in response to an exposure.

• There is reason to believe that exposure is possible or likely.

On the basis of those characteristics, several criteria and methods for selecting target chemicals are possible. A recent discussion of priorities for laboratory testing of chemicals (NRC, 1984a) presented many issues that are relevant here. They are now briefly described.

Selection Based on Analytic Expediency and Constraints

In formulating a new monitoring program, one might choose analytic methods to maximize the likelihood of detection of key target chemicals. Multiagent analytic schemes that rely on gas chromatography and mass spectrometry (GC-MS) are common to many environmental monitoring laboratories. Within any such scheme, the ranges of applicability and sensitivity of detection are roughly fixed; analytical performance then dictates what goals (i.e., what list of target chemicals and detection limits) might be considered. For example, schemes currently used to detect EPA priority pollutants with detection limits of 1-10 ppb in water or wastewater might permit inclusion of additional target chemicals with similar analytic behavior (for example, alkylated polycyclic hydrocarbons in addition to the priority polycyclic aromatic hydrocarbons), but could not permit inclusion of polychlorinated dibenzodioxins (PCDDs) and dibenzofurans (PCDFs) at concentrations lower than parts per billion. Modification of analytic protocols might permit detection of PCDDs and PCDFs, but exclude detection of polycyclic hydrocarbons altogether.

For monitoring purposes, established assay protocols represent analytic "windows" through which specific subsets of contaminant chemicals might be detected and measured. That is an advantage, because within such a window nontarget chemicals might be detected and new target chemicals added without extensive analytic-method development; but it is a disadvantage because "blind spots" outside established analytic windows might be neglected. The latter concern would be reinforced if analytic expediency were a main criterion for identifying new target chemicals.

Selection Based on Regulatory Interest

To the degree that a monitoring program is aimed at a specific regulatory issue, other criteria for the selection of target chemicals might be preempted. Analytic methods would then have to be devised to meet the requirements of the chemicals targeted by the regulation, and future additions to the list would probably require development of new methods. It is important to determine what regulatory needs are to be served by a monitoring program, but it is equally important to determine whether the program is confined to those particular needs or has (or will have) a broader mission as well.

Selection Based on
Prior Detection in Human Tissues

Reports of detection of contaminant chemicals in human tissue in the scientific literature or in regulatory-agency studies constitute convincing evidence that exposures can occur and that tissue concentrations are measurable, by at least some method. Compilations of such reports—as in the Chemicals Identified in Human Media data base (EPA, 1980) and the NIST Human Specimen Banking Reports (Wise and Zeisler, 1984)—not only are tools for locating candidate target chemicals that might conform to pre-existing analytic capabilities, but also might indicate a need for new analytic approaches to widen the range of chemicals detected.

Selection Based on
Indications of Health Relevance

Testing programs conducted by NIH or by nongovernment researchers will continue to provide findings that increase or decrease health concerns related to possible exposures to individual agents. In lieu of case reports or other documented findings of agents in tissues to be monitored, results of toxicologic studies in animal models can indicate the likelihood that a candidate target chemical will be present and detectable in sample tissues. When such findings indicate a need for epidemiologic followup, inclusion of the agent in monitoring programs should be considered.

Selection Based on Indications
of Environmental Contamination

Numerous environmental monitoring programs focus on drinking water (Wallace et al., 1986), air quality (Hunt et al., 1986; Wallace et al., 1986; EPA, 1987b), food (Reed, 1985; Reed, et al., 1987), biota (Lewis and Lewis, 1979; Becker et al., 1988; NOAA, 1988), and waste treatment and discharges (Hannah and Rossman, 1982). Findings that indicate widespread environmental contamination by specific chemicals would add to the importance of including those chemicals in a program of tissue monitoring.

Summary

Clearly, those approaches are not independent. For example, regulatory priorities reflect toxicologic research, environmental monitoring, etc., and it is probable that some candidate target chemicals will satisfy several criteria. But, each approach can identify agents not currently being studied or considered in connection with the other approaches.

Design of Chemical Assay Programs

Determination of Requirements for Analytic Performance

Precision, accuracy, and sensitivity must all be adequate to meet the specified goals of the program. Analytic goals are ideally based on knowledge of the relationships among tissue concentration, exposures, and health outcomes and on the variations in tissues concentrations that might be attributed to individual variations, to systematic variations in exposure, and to the time course of exposure. Such information is rarely available in practice, but analytic precision, sensitivity, and accuracy should be great enough to permit detection and measurement of exposures well below any plausible threshold of clinically evident effects.

Decisions regarding type of detection method to be used or whether multiple target chemicals are to be included in a single assay (and if so, which ones) will depend on the analytic performance required. Specificity and sensitivity are closely related; when analytic needs are defined on the basis of high unit toxicity or a need to detect an agent in tissues of persons exposed at background concentrations, a dedicated assay may be required for the specific agent or a group of agents.

Development and Validation of Method

Ways of developing and validating specific analytic techniques should be specified in considerable detail and then applied to each chemical in the initial set or added later. At a minimum, method development will include modification of existing protocols and methods developed for use in other applications or taken from the scientific literature, so that one can demonstrate adequate performance for the specific target chemicals, tissue sample type, and concentration range desired. Some of the steps (not necessarily in order) are:

- Demonstration that the method is adequately sensitive to detect each target chemical by assay of appropriate coded (blind) standards.
- Demonstration of adequate recovery of target chemicals in sample preparation and instrumental assay steps through analysis of standard mixtures in a blank (matrix-free) solution.
- Demonstration of adequate control of interference by chemicals in the sample background through analysis of both blank samples and spiked matrix samples.
- Demonstration of good overall recovery of target chemicals from representative samples. Actual (unfortified) samples that have been assayed thoroughly are needed for this step, because biologically incorporated target chemicals might not be as readily isolated from the sample as would a spiked contaminant.

Several of those steps should be performed in replicates, so that assay variability can be assessed. More extensive method development will be required for analytic needs or target chemicals that do not conform closely to an existing protocol. If new or improved separation steps or methods of detection are needed, they must be developed before the steps outlined above.

Once a method is demonstrated to meet stipulated analytical requirements, a validation study should establish the performance of the method with real samples over a realistic period and, if possible, with a range of operators, equipment, and laboratories. Precision would be characterized with replicate and coded (blind) analysis or assay of control materials, if available. Method accuracy would be established with reference samples or interlaboratory comparisons.

Completion of this formalized method development and performance evaluation is necessary to support quantitative uses of monitoring results. Without such knowledge of system recoveries and performance, tissue concentrations cannot be defined and negative results cannot be properly interpreted.

Design of Quality Assurance

A detailed quality-assurance (QA) plan should formalize the analytic procedures for each assay. If results are to be acceptable it is necessary to specify calibration materials, acceptable levels for blanks, recovery samples, replicate analyses, other control samples, and the frequency with which procedures are applied. QA plans can also specify preventive-maintenance schedules for instruments, how data are to be validated, and what remedial procedures are to be used when QA procedures detect unacceptable results. The overall intent is to build into the assay a system of diagnostic measurements that detect and correct degradation of analytic performance. Two tools of particular value in maintaining long-term analytic stability are the benchmark sample and the graphic control chart.

The benchmark sample is of the same type as study samples and ideally has detectable concentrations of several or all of the target chemicals. It is of known stability and is prepared in many multiple replicate subsamples. The chemical homogeneity of these subsamples is established by assay of a statistical fraction of the total number of aliquots prepared. The sample might be composited from many individual samples, provided that the blending process achieves adequate homogeneity. Analysis of portions of the sample over time is used to detect and document changes in assay response. Correlation of assay results with the benchmark sample among methods and laboratories is used to establish reference values for the contaminants in the sample, which make it possible to verify assay accuracy and to detect assay bias.

After statistical limits are established for subsample homogeneity and assay precision in the benchmark sample (usually as a part of validation experiments), a control chart is established for individual target chemicals. Results of reanalysis of the benchmark sample are posted on this chart, and control limits (usually reflecting 95% confidence values from the validation data set) are established. Maintenance of a correct control chart allows assay acceptability to be checked on a day by day or even batch by batch (Taylor, 1985).

Identification of Additional Resources Needed

Although it is possible for any monitoring program to prepare control materials by blending a large volume of samples, an alternative is to use standard reference materials provided by NIST and other organizations. Reference materials offer several advantages over blended samples: stability and homogeneity have been well established; reference values for certified contaminants are well established and documented; and a large community of users is sharing the same materials, so that evaluation of comparability of results is enhanced. The appropriateness of existing reference materials to the needs of any monitoring program will depend on the match between matrix type, analyte list, and concentration range in standard reference materials and monitoring-program samples and target chemicals.

For a monitoring program that will continue for decades, development of reference materials should have high priority. Not only will such materials help in maintaining and documenting consistent monitoring results, but provision of portions of the materials to researchers or other monitoring programs will help in establishing analytic comparability for related studies.

Among the kinds of additional resource needs that might be identified at this point are alternative methods for specific analytes that show marginal or unacceptable analytic performance; identification of intermittent contamination sources, such as particular types of containers or sampling media; and data on the chemical stability of components that display high variability of recovery.

Implementation of Method
Development and Validation Studies

Analytic-method development often involves complex choices, particularly when many kinds of performance must be optimized in a single procedure. If the choices are to be optimal, an analyst who focuses on specific needs or problems in the assay scheme must work closely with the program planner who is considering future needs and larger objectives. In general terms, the process of preparing a sample for assay can be viewed as a series of steps that exclude chemical constituents from the sample, simplify the sample composition, and increase the ability to detect and measure the target chemicals that remain. Collection of components from the sample by extraction or other phase-separation techniques is the first stage. In the optimization of a method, there are commonly many solutions to a given problem. Substitution of one chromatographic medium for another and modification of the mobile phases used to partition sample components in prefractionation steps are two examples. Given a choice between two or more procedures that accomplish the basic task of removing interfering (and presumably unimportant) sample constituents, the alternative that best meets other program needs (such as a high likelihood of including agents of possible future monitoring interest) should be selected. Such an alternative should, of course, be validated in practice before it becomes a part of routine operations.

Pilot-Scale Monitoring

Pilot-scale testing will identify and help to correct problems in the various elements of a monitoring program, including collection and transmission of samples, sample management before assay, chemical measurements, data management, and reporting. Each modification of the assay opens new possibilities for unforeseen technical problems. For example, addition of new target chemicals to an otherwise unchanged assay might require alteration of materials used for sample collection, to avoid contaminating a sample with a new target chemical. In particular, when a new assay is added to a program, pilot testing of that assay might prevent large-scale waste of effort and re-sources.

Continuing Monitoring:
Review of Analytic Program and Goals

The QA plan should contain a statement of what constitutes acceptable data and acceptable assay performance. A schedule of data review and valida-tion, with periodic QA review, should be created. Outside review of the QA plan itself is desirable, as is periodic auditing of quality control-procedures and results by a group separate from the monitoring program. The standing committee that we recommend elsewhere in this report is a possibility.

Monitoring of assay procedures will satisfy program objectives that are based on a specific monitoring need and a defined list of agents of concern. There will be incentives for further development within the narrowly defined mission to measure the same target chemicals more accurately, more quickly, and more cheaply. Not only can improved measurement technology permit dramatic increases in sensitivity and ease of measurement, but failure to keep pace with the state of practice will eventually weaken the credibility of moni-toring-program results. Monitoring-program compromises imposed by analyt-ic limitations, such as the number of assays supportable by program budgets or the requirement for compositing to ensure assay sensitivity, might be re-duced or eliminated as technology improves.

The desire to extend the monitoring program to new agents and to recog-nize otherwise unanticipated exposures is a second reason to encourage devel-opmental tasks within a monitoring program. Reasons for incorporating new agents into a national monitoring program are to respond to needs and con-cerns broader than individual regulatory programs and to realize additional benefits from the considerable resources that a monitoring program requires. Exploratory activities within a monitoring program can provide an anticipatory approach to hazard recognition and provide for efficient application of new findings in the basic program. Incorporation of new agents into existing assay protocols and development of new protocols to address new monitoring goals can be planned as continuing activities.

Exploratory Analyses

Exploratory analyses are of two sorts: those that attempt to catalog nontarget contaminants as they occur in monitoring of sample preparations and those that seek to detect classes of chemicals that are not expected to occur in significant concentrations in assay-sample fractions but would be of great interest if they did. The latter kind might use sample-preparation procedures that differ from assay procedures or instrumental detection techniques other than the assay techniques used in monitoring. Results will be qualitative or at best semiquantitative, because method recoveries and other performance would not have been determined and instrument response calibration with valid standards might not have been possible.

"Volunteer" Chemicals in Monitoring Assays

Modern assay techniques, such as GC-MS, permit a degree of qualitative analysis of sample constituents other than target chemicals. Mass spectrometry as a detection and quantitation method typically involves collection and storage of sequential mass spectra representing the entire detectable portion of the sample or solution analyzed. Computerized comparisons with mass-spectrum libraries can tentatively identify some proportion of "unknown" components, if the library of standards contains mass spectra similar to the unknown spectrum from the sample. The reliability of a tentative identification of an unknown component will depend on the intensity and purity of the unknown spectrum and on the nature of the chemical detected. Confirmation of tentative identifications with standards to demonstrate matching mass spectral and gas chromatographic behavior is required. Alternatively, when sufficient quantities of the unknown can be isolated from a sample, the identities of the unknown might be determined by mass-spectrum interpretation combined with traditional techniques for the elucidation of chemical structure (infrared, and ultraviolet spectroscopy, nuclear magnetic resonance, elemental analysis, and high-resolution mass spectrometry). The undertaking is time-consuming and expensive, so a standardized approach might be desirable for deciding which unknowns should be identified. Some considerations might be apparent concentration (i.e., intensity of GC-MS peak); frequency of detection; mass-spectral features that suggest anthropogenic origins such as patterns indicative of chlorine or bromine atoms; or similarity of mass spectra to those of known toxicants.

Alternative Detection Methods

The versatility and specificity of GC-MS make it the most widely used tool for environmental monitoring of organic contaminants in environmental samples. Under electron-impact ionization conditions, it is possible to generate similar mass spectra from a variety of instrument designs, so large public-domain mass-spectrum libraries can be used. The adoption of GC-MS by EPA for many of its monitoring methods has led to further standardization of MS instrumental characteristics. However, several techniques are important complements to standard GC-MS. Chemical ionization improves detection of chemicals that show excessive fragmentation under electron-impact ionization conditions. Both positive and negative chemical ionization can also be used selectively to improve detection of chemicals with higher gas-phase proton or electron affinities, as well as providing molecular-weight information. Although those techniques are not widely used for routine monitoring, they are particularly useful for detecting traces of halogenated chemicals, such as PCDDs (Dougherty et al., 1980).

Other emerging techniques can address those chemicals that are not detectable by any gas chromatographic technique because they are nonvolatile or thermally unstable. Among those methods are new interface designs to permit high-performance liquid chromatography and mass spectrometry (LC-MS) (Covey et al., 1986) and supercritical-fluid chromatography and mass spectrometry (SFC-MS) (Smith et al., 1986). Secondary-ion mass spectrometry with collisional activation, an MS-MS technique (Tondeur et al., 1987), is another approach that permits detection, identification, and ultimately quantitation of nonvolatile chemicals at low concentrations in biologic samples. Each of those techniques could play a useful role in exploratory analyses, and their incorporation into monitoring protocols will soon be feasible. The newer techniques open the way for detection of polar metabolites of toxic agents, as well as of "refractory" toxicants. Innovative uses of new approaches to develop new information for human-tissue monitoring should be encouraged.

Alternative Sample-Preparation Schemes

Exploration of alternative sample extraction and prefractionation techniques should occur as new analytic techniques are developed. "Throwaway" fractions obtained from monitoring protocols that are not amenable to conventional GC-MS analysis could be screened with LC-MS, SFC-MS, and MS-MS. Supercritical-fluid extraction methods (Kalinoski et al., 1986) and microcolumn liquid-chromatographic methods (Ozretich and Schroeder, 1986) are receiving wide attention for use in sample preparation.

Establishing Goals for New Assay Technology

It is critically important in selecting exploratory projects to consider all available information that bears on the presence and metabolic fate of toxicants. Without clear hypotheses regarding possible positive findings, exploratory analyses become "fishing expeditions" with little likelihood of producing useful results. Close coordination with environmental monitoring programs, as well as with bioassay and other toxicity-testing programs, is essential. Program design in this field will need to rely on scientific advisers who are familiar with the state of analysis and toxicology and who can realistically assess the probable benefit of a given exploratory program.

PRACTICAL LIMITATIONS AND COMPROMISES

Several kinds of compromises can reduce overall costs associated with a monitoring program. Some reduce development costs and lead time, some reduce the number of assays needed for monitoring, and some reduce the cost per assay. Reductions in development costs can be achieved by targeting analytes with similar chemical properties to minimize the number or complexity of analytic protocols or by combining method development with actual sample analysis. Both approaches have some merit, although as noted above, they also have disadvantages and risks.

Methods to Reduce the Number of Analyses

Compositing of Samples

According to the present analytic schemes for extractable semivolatile chemicals and dioxins and furans, there is no compelling need to composite individual specimens. (The NHATS's decision to composite samples was driven by cost, not sensitivity.) Current detection limits are in the low nanograms-per-gram range for semivolatile chemicals and the low picograms-per-gram range for dioxin and furans for assay of a 1% aliquot of a 20 gram tissue sample (semivolatile chemicals) or a 10% aliquot of a 10-g tissue sample (dioxins and furans). Revision of sample-size standards, fraction used per assay, and target-tissue collection amount could permit analysis of individual samples in most cases. As a reason for compositing samples, cost must be balanced against the statistical limitations imposed by compositing.

Two situations in which compositing is reasonable are method-development studies (where a feasibility-study scale is preferred to a full scale) and qualitative and exploratory assays, where larger tissue pools are needed and where the cost per sample is much higher than for monitoring analysis. Composite samples are needed in developing large, uniform quantities of standard reference materials.

Subsampling or Stratified Assays

An alternative to compositing is subsampling; individual samples are selected from a larger pool. The issues in selecting a subsample are primarily statistical; however, limited chemical analyses might be used to select tissues with substantial burdens of a few chemicals that may be markers for other exposures. "More exposed" and "less exposed" persons might be identified for subsample analysis, on the basis of hypothesized or demonstrated relationships between the results of detailed monitoring and the results of use of indicator chemicals.

Less-Frequent Analysis or Staggered Assays

The schedule of analysis, the frequency of reporting, and the overall delay from specimen collection to publication of results need special attention in any monitoring program. Too-frequent reporting will result in unnecessary costs and perhaps obscure critical findings, because the amount of new data in each report will be small. Delay of publication, whatever the reasons, diminishes the value of the data, and the loss of value increases with increasing delay, even to the point where findings can be of little current interest. A staggered schedule for statistical analysis and reporting can break the overall workload into more manageable tasks, but at the cost of a larger administrative burden (more reports), the loss of some information on joint exposure, and a loss of attention in the user community.

Special Studies

The characterization of particular exposure situations, such as a geographic region of high environmental contamination or those that focus on special subpopulations, is more efficiently addressed by special studies than by a nationwide survey. Special studies are crucial to the NHMP goal of establishing links among quantitative environmental contamination, specific targeted exposures and tissue concentrations, and the goal of detecting new exposures. EPA has used special studies to respond to interests and needs outside the NHMP itself. However, one way for EPA to reduce analytic costs while meeting its stated program goals for national monitoring is to use special studies in conjunction with other cost-reduction strategies such as less-expensive analyses, shortened target-chemical lists, and methods with less-than-maximal sensitivity. FDA's "sensitivity of the method" approach—that without human data, data from the most sensitive species should be considered to be most like data that might be gathered from humans—established a regulatory precedent, but that option runs counter to the 1982 decision to extend and modernize the chemical analyses in the NHATS program. Furthermore, NHATS uses multichemical analytic schemes, and the incremental savings achieved by deleting one or a few analytes is small. It seems inappropriate to raise any current detection limit requirements, because few agents are routinely detected at concentrations that would be reached with less-sensitive analysis.

CURRENT AND PREVIOUS ANALYTIC PRACTICES
OF THE NHATS PROGRAM

The chemical agents listed in Table 6-1 as fiscal year 1982 target chemicals were intended to cover anthropogenic classes that had been reported in published analyses of human adipose tissue, including phthalate esters, phosphate triesters, polychlorinated aromatics, polychloroterphenyls, polybrominated biphenyls, and polycyclic aromatic hydrocarbons, in addition to the organochlorine pesticides and PCBs already monitored. In general, those chemicals either have been reported to occur in adipose tissue or could be expected to accumulate there, because of their lipophilicity and (in some cases) resistance to metabolic conversion to less lipophilic products.

Several ways were used to nominate new target chemicals. Although some basis for their selection can be inferred from the EPA literature supplied to the committee, the rationale is not clear. For example, the inclusion of brominated dibenzodioxins and dibenzofurans as target chemicals without method validation in the planned analysis of 1986 specimens was based not on prior detection in adipose tissue, but on "the potential for exposure to brominated analogues of dioxins and furans from specific brominated commercial products" (MRI, 1988).

Another gap in the apparent rationale is the inclusion of additional "one-time" efforts to survey elemental composition and volatile organic compounds in 1982 adipose-tissue samples. Distribution into adipose tissues is known to

TABLE 6-1 NHATS Target Compounds

I. Target compounds:
semivolatile analysis protocol

Compound	CAS No.	History or Source[a]						
		A	B	C	D	E	F	G
Pesticides								
Lindane (γ-BHC)	58-89-9	x	x			x		
Mirex	2385-85-58	x				x		
Chlordane	57-74-9	x	x			x		
Heptachlor	76-44-8	x	x			x		
Endrin	7221-93-4	x				x		
Endrin ketone		x		x		x		
Dieldrin	60-57-1	x		x		x	x	
p,p'-DDT	50-29-3	x				x		
Aldrin	309-00-2	x	x	x		x		
α-BHC	319-84-6	x		x		x		
6-BHC	319-85-7	x		x		x		
δ-BHC	319-86-8	x		x		x		
Heptachlor expoxide	1024-57-3	x				x		
t-Nonachlor	39765-80-5	x				x		
o,p'-DDT	789-02-6	x				x		
o,p'-DDE	3424-82-6	x				x		
p,p'-DDE	72-55-9	x				x	x	
o,p'-DDD	53-19-0	x				x		
Oxychlorodane	26880-48-8	x				x	x	

Compound	CAS No.	History or Source[a]						
		A	B	C	D	E	F	G
Isophorone[b]	78-59-1			x	x			
Dicofol[b]	115-32-2						x	
Dichlorvos[b]	62-73-7		x				x	
Methoxychlor[b]	72-43-5		x				x	
Nitrofen[b]	1836-75-5		x				x	
Butachlor[b]	23184-66-9						x	
Chlorpyrifos[c]	2921-88-2						x	
Isopropalin[c]	33820-53-0						x	
Perthane[c]	72-56-0						x	
Trichloronate[c]	327-98-0						x	
Chlorinated Aromatics								
1,3-Dichlorobenzene	541-73-1		x	x	x	x		
1,4-Dichlorobenzene	106-46-7		x	x	x	x		
1,2-Dichlorobenzene	95-50-1		x	x	x	x		
1,2,4-Trichlorobenzene	120-82-1		x	x	x	x		
Hexachlorobenzene	118-74-1	x	x	x	x	x		
1,2,3-Trichlorobenzene	87-61-6					x	x	
1,3,5-Trichlorobenzene	108-70-3					x	x	
1,2,3,4-Tetrachlorobenzene	634-66-2					x	x	
1,2,3,5-Tetrachlorobenzene	634-90-2					x	x	
1,2,4,5-Tetrachlorobenzene	95-44-3					x	x	
Pentachlorobenzene	608-93-5					x	x	

Compound	CAS					
Hexachloronaphthalene[c]	1335-87-1	x				
Octachloronaphthalene[c]	2234-13-1	x				
Pentachloroanisole[c]						
Pentachloronitrobenzene	82-68-8					x
2,3,6-Trichlorophenol[c]	933-75-5					x
2,4,5-Trichlorophenol[c]	95-95-4					x
2,4,6-Trichlorophenol[c]	88-06-2					x
2,3,6-Trichloroanisole[c]	50375-10-5					x
2,3,5-Trichloroanisole[c]	54135-80-7					x
2,4,6-Trichloroanisole[c]	87-40-1					x

Polynuclear aromatic hydrocarbons (PAHs)

Compound	CAS					
Napthalene	91-20-3	x	x	x	x	
Phenanthrene	85-01-8		x	x	x	
Flouranthrene	206-44-0		x	x	x	
Chrysene	218-01-9		x	x		
Benzo*a*pyrene	50-32-8			x		
Acenaphthylene	208-96-8					x
Acenaphthene	83-32-9					x
Flourene	86-73-7					x
Phenathrene	85-01-8					
Pyrene	129-00-0					x

Compound	CAS No.	History or Source[a]						
		A	B	C	D	E	F	G
Anthracene[c]	120-12-7		x		x			
Benzoaanthracene[c]	56-55-3		x	x		x		
Benzoaflouranthene[c]	205-99-2		x	x		x		
Dibenzoa,hanthracene[c]	53-70-2		x	x		x		
Indeno1,2,3-cdpyrene[c]	193-39-5		x	x		x		
Benzokflouranthene[c]	207-08-9					x		
PCBs[b]								
Monochlorobiphenyl	2735-18-8		x	x		x	x	
Dichlorobiphenyl	25512-42-9		x	x		x	x	
Trichlorobiphenyl	25323-68-6		x	x		x	x	
Tetrachlorobiphenyl	26914-33-0		x	x		x	x	
Pentachlorobiphenyl	25429-29-2		x	x		x	x	
Hexachlorobiphenyl	26601-64-9		x	x		x	x	
Heptachlorobiphenyl	28655-71-2		x	x		x	x	
Octachlorobiphenyl	314-83-0		x	x		x	x	
Nonachlorobiphenyl	53742-07-7		x	x		x	x	
Decachlorobiphenyl	2051-24-3		x	x		x	x	

Phthalate esters

Compound	CAS				
Dimethyl phthalate	131-11-3		x	x		
Dibutyl phthalate	84-74-2		x	x		
Butylbenzyl phthalate	85-68-7		x		x	
Di-*n*-octyl phthalate	117-84-0		x	x	x	
Diethyl phthalate	84-66-2		x			x
Di-*n*-butyl phthalate	84-74-2		x			x
Diethylhexyl phthalate (DEHP)	117-81-7		x			x

Phosphate triesters

Compound	CAS				
Tributylphosphate	126-73-8					
Triphenylphosphate	115-86-6		x			x
Tris(2-chloroethyl)phosphate	115-96-8		x			x
Tributoxyethylphosphate	78-51-3		x			
Tritolylphosphate	1330-78-5		x			
Tris(dichloropropyl)phosphate	n/a					
Tris(2,3-dibromopropyl)-phosphate	126-72-7	x				x

History or Source[a]

Compound	CAS No.	A	B	C	D	E	F	G
Other								
1,2-Dibromo-3-chloropropane	96-12-8		x		x			
Hexachloro-1,3-butadiene (HCBD)	87-68-3		x	x	x			
Hexachlorocyclopentadiene	77-47-4		x		x			
Biphenyl	92-52-4		x		x			
2-Phenylphenol	90-43-7		x					x
4-Chloro-p-terphenyl						x		
Tetrabromobiphenyl						x		
Pentachlorodiphenyl ether						x		
o-Cymene (o-isopropyl toluene)	527-84-4							x
m-Cymene (m-isopropyl toluene)	535-77-3						x	
p-Cymene (p-isopropyl toluene)	99-87-6							
D-Limonene	5898-27-5							x
DL-Isoborneol	124-76-5							x

Compound	CAS Number				
1-Indanone	83-33-0				
2-Indanone	615-13-4				
2-Methoxy-3-methylpyrazine	2847-30-5				
Butylated Hydroxytoluene	128-37-0				
Ethyl hydrocinnamate	2021-28-5				
Coumarin	91-64-5				
Octamethyl cyclotetrasiloxane	556-67-2				
Cumene[c]	98-82-8	x			
1,2,4-Trimethylbenzene[c]	95-63-6	x			
Quinoline[c]	91-22-5	x			
Dibenzofuran[c]	132-64-9	x	x		
Chlorbenzilate[c]	510-15-6			x	
Bis(2-ethylhexyl)adipate[c]	103-23-1	x			
o-Xylene[c]	95-47-6				x
m-Xylene[c]	141-93-5				x
p-Xylene[c]	105-05-5				x
2-Ethoxy benzaldehyde[c]	613-69-4				x
1-Nonene[c]	124-11-8				x
1-Pentanol[c]	71-41-0				x
Hexyl acetate[c]	142-92-7				x

Compound	CAS No.	History or Source[a]						
		A	B	C	D	E	F	G
II. Target Compounds: PCDD, PCDF Protocol								
Chlorinated dibenzodioxins and dibenzofurans								
2,3,7,8-tetrachlorodibenzo-dioxin						x		
1,2,3,7,8-pentachlorodibenzo-dioxin						x		
1,2,3,4,7,8-hexachlorodibenzo-dioxin						x		
1,2,3,6,7,8-hexachlorodibenzo-dioxin						x		
1,2,3,7,8,9-hexachlorodibenzo-dioxin					x			
1,2,3,4,6,7,8-heptachlorodi-benzodioxin						x		
Octachlorodibenzodioxin						x		
2,3,7,8-tetrachlorodibenzo-furan						x		
1,2,3,7,8-pentachlorodibenzo-furan						x		
2,3,4,7,8-pentachlorodibenzo-furan								

1,2,3,4,7,8-hexachlorodibenzo-furan		x
1,2,3,6,7,8-hexachlorodibenzo-furan		x
1,2,3,7,8,9-hexachlorodibenzo-furan		x
2,3,4,6,7,8-hexachlorodibenzo-furan		x
1,2,3,4,6,7,8-heptachlorodi-benzofuran	x	
1,2,3,4,7,8,9-heptachlorodi-benzofuran	x	
Octachlorodibenzofuran		x

Polybrominated dibenzodioxins and dibenzofurans
2,3,7,8-tetrabromodibenzo-dioxin[d]
1,2,3,7,8-pentabromodibenzo-dioxin[d]

History or Source[a]

Compound	CAS No.	A	B	C	D	E	F	G
1,2,3,4,7,8-hexabromodibenzo-dioxin[d]								
2,3,7,8-tetrabromodibenzo-furan[d]								
1,2,3,7,8-pentabromodibenzo-furan[d]								
1,2,3,4,7,8-hexabromodibenzo-furan[d]								

[a] A: Historical NHATS pesticide analyte (1970-1981)
B: SARA 313 chemical list
C: SARA/ATSDR 110 chemical list
D: EPA Method 1625 analyte
E: Target analyte in analysis of fiscal year 1982 adipose tissue samples
F: Bioaccumulative pollutant study target analyte, EPA contract 68-01-6951, WA7 and 13
G: Analytes previously qualitatively identified in broad scan analysis of fiscal year 1982 specimens

[b] Individual congeners used as standards; detected as mixtures in historical NHATS surveys
[c] Nonvalidated target analytes for qualitative analysis
[d] Nonvalidated target analytes for quantitative analysis

play an important short-term role in the toxicokinetic behavior of many common volatile solvents, but they are not known to accumulate in adipose tissue over long periods. The methods used for sample collection and storage had not been validated for volatile chemicals, and the committee questions the integrity of the samples for those applications. Little information exists on the concentrations of trace elements in adipose tissues from normal, abnormal, or overexposed persons, and the diagnostic value of such measurements is uncertain. The assays are not planned for incorporation into routine monitoring of adipose tissues.

Selection of analytic methods was generally based on methods already used and reported in published applications to analysis of adipose tissues (Macleod et al., 1982), other tissues (Norstrom et al., 1986), and environmental samples (Lopez-Avila et al., 1981). The general NHATS approach consists of exhaustive extraction of tissue with a lipid solvent such as methylene chloride; prefractionation of the extract with size-exclusion chromatography to remove the bulk of biologic (predominantly lipid) background material from the extract; additional prefractionation with one or more conventional (adsorption-partition) column chromatographic steps to separate out the least polar (aliphatic) components and to separate interfering groups of target chemicals (e.g., toxaphene or other chemicals from PCBs); and final instrumental detection and quantitation with high-resolution GC/low-resolution MS. The scheme for sample analysis as presented generally in the 1989 "Program Strategy" document and more specifically in the reports resulting from the initial applications of those methods to adipose samples (Mack and Stanley, 1984) is shown in Table 6-2.

The analysis of 1982 samples according to the new methods took place in 1984-1985 and was reported in 1986. Additional validation and method-development studies were undertaken concurrently with and after the 1982 tissue analyses. All four of the new analytic projects (semivolatile chemicals, dioxins and furans, trace elements, and volatile chemicals) are most properly regarded primarily as method-development exercises and only secondarily as tissue surveys. Because of the lack of prior method validation, the results of those analyses are referred to in the respective project reports as concentration estimates.

An intermethod comparability study that used historical and 1984 samples was initiated to compare the historical pesticide-survey method with the pesticide results obtained under the 1982 protocol. That effort was carried out at a contracting laboratory other than the one that had performed the 1982 method development and analysis. Some problems in applying the newer protocols were reported, and results of the study are not yet released. Additional method development and application of the semi volatile-chemical

TABLE 6-2 NHATS Analytic Efforts

Collection Year	Activity	Analytes	Assay Method	Completion Date or Status
1970-1981	Monitoring	OCl pesticides	PGC/ECD[a]	Annual
1982	R&D, survey	OCl pesticides, semivolatiles	HRGC/MS[b]	1986
	R&D, survey	Volatiles	HRCG/MS[c]	1986
	R&D, survey	Trace elements	ICP, NAA[d]	1986
	Validation	PCDDs, PCDFs	HRGC, MS[e]	1986
	R&D, survey	PCDDs, PCDFs	HRCG/MS[b]	1986
	Exploration	Unknown peaks	HRGC/MS[f]	1986
1983	Monitoring	OCl pesticides	PGC/ECD[a]	Partially published
1984	Comparability study	OCl pesticides, semivolatiles	PGC/ECD[a] vs. HRCG/MS[b]	Under review
1985	No analyses	n/a	n/a	n/a
1986	Monitoring	OCl pesticides, semivolatiles	HRGC/MS[b]	In progress
	Exploratory	Unknown peaks	HRGC/MS[f]	In progress
1987	Monitoring	PCDDs, PCDFs	HRGC/MS[b]	Under review
	R&D, survey	PBDDs, PBDFs	HRGC/MS[b]	In progress
1988	Analyses being planned	n/a	n/a	n/a

[a]Modified Mills-Gaither-Olney pesticide method with low-resolution gas chromatography with electron-capture detection.

[b]Extraction, size exclusion, and adsorption chromatographic sample prefractionation with high-resolution gas chromatography and mass-spectrometry detection.

[c]Purge and trap collection of vapors from an aqueous slurry of tissue sample with high-resolution gas chromatography and mass spectrometry detection.

[d]Acid sample digestion and quantitation with atomic-absorption spectrometry or inductively coupled plasma atomic-emission spectroscopy.

[e]Solid-phase extraction and digestion of lipid matrix, adsorption chromatographic prefractionation on graphitized carbon, with high-resolution gas chromatography and mass-spectrometry detection and high-resolution mass-spectrometry confirmation of key analytes.

[f]Extraction, size exclusion, and adsorption chromatographic sample prefractionation with high-resolution gas chromatography/mass spectrometry detection; unknown peak screening by mass spectral library comparison to reference spectra and to previously observed unknown components.

("broad-scan") protocol has occurred since then, but no formal results have been released.

The target chemicals listed in part I of Table 6-1 were taken from the QA planning document for the most recent round of analysis (1986 sample year). Target chemicals designated with superscripts c and d are planned for inclusion as "qualitative" or "quantitative" analytes. Method validation has not been performed, but these chemicals are detectable with the GC-MS method to be used, and standards are available.

The most successful of the method-development and validation projects, and one that resulted in the production of a well-documented protocol as a basis for future monitoring, was the dioxin-furan project. Several documents presented the accomplishments in PCDD and PCDF analysis. The first is a report of an analysis of 1982 tissue-survey samples that used developmental methods. The analytic work took place from the fall of 1984 through the winter of 1985 (EPA, 1986a). The report reflects an improvisational approach to the analytic method that is incompatible with monitoring goals, because EPA, through the NHATS, had attempted to implement methods that were not fully tested and validated. One must infer which analysis produced which results by examining the size of the tissue aliquot reported (Tables 5-14 in that report). For given target chemicals, different samples were assayed with different methods. Some samples were assayed with two methods, and that would presumably produce results for the complete list of target chemicals for each method, but the results are reported for one method for some analytes and for the other method for other analytes. Comparison of results of the two methods is limited to mean and standard deviation of recoveries for the two sets of data. No side-by-side comparison of results for each preparation method is shown, although such data were presumably generated. The consequences of that approach include the use of multiple methods of sample preparation when the method initially adopted was judged to be inadequate for some samples of target chemicals, lack of clarity about which methods produce which results, and uncertainty regarding comparability of results of the two methods. Method-related uncertainties, and a generally tentative level of confidence in the quality of the analysis are reflected in the following language of the report: "The data for a sample reported based on the original protocol may be considered suspect based on the possible differences in the recoveries of these chemicals according to the two methods used". "A continued effort in following the trends of PCDD and PCDF will require that the analytical method . . . be fully validated through intra- and interlaboratory studies."

The second report describes method-development and validation work that was largely after the initial application of the methods to study samples (EPA,

1986b). The effort addressed several of the problems of the previous results. The report is a good model for future method-development and validation projects. It discusses several prerequisites of the production of fully usable data: method precision, accuracy, sensitivity, and stability are characterized in a statistical manner; control materials and comparison results are developed and validated; and a step-by-step approach to quality control throughout the analytic protocol is developed.

The revision of the latter project report contained in the third document indicates an intention to use the control material developed previously as a continuing quality control element (MRI, 1988). Additional planning for sample and data management and quality control are also reported.

Preliminary efforts like those reported in the second and third documents provide a good basis for a monitoring effort. However inclusion of new but related chemicals in the overall framework of such a program introduces new risks. For example, the proposed inclusion of brominated analogues of PCDD and PCDFs as target chemicals without an explicit validation effort is a short cut that can lead to "messy" and unsatisfactory results. Overall confidence in the results of the program would be strengthened if results of "try it and see" assays of study samples were more clearly distinguished from results of well-validated analyses.

The NHATS or its successor must document the rationale and specific analysis that result in program decisions during each year's planning to add or delete target chemicals.

SUMMARY AND RECOMMENDATIONS

The NHATs project reports show several encouraging characteristics. Some of the weaknesses identified and discussed in the previous section have been recognized by EPA and its contractors, and efforts to remedy them are evident. Within the definition of each analytic task, the contractor has demonstrated a competent and sometimes innovative approach to analytic-method development. The most recent reports and planning documents show a good understanding of quality-assurance planning and quality-control techniques. In summary, there is no reason to doubt that the program can attain state-of-the-art analysis. The principal concerns in this regard would be with constraints imposed on the analytic effort by inadequacies of agency planning and direction, budgeting, or long-term commitment to a program of human-tissue monitoring.

Several factors make the need for analysis of program goals and a sound planning process critical: the need to balance innovative method development with the maintenance of stable assays for monitoring; the open-ended nature of the analytic effort and the cost per sample; the need for sufficient analyses to achieve sampling and statistical goals; and the inherently multiagency and multiuse nature of the program, which requires strong coordination among agencies and with other sources of technical information vital to program planning.

Definition of Goals

Efforts to date have been oriented toward those chemicals that previous reports suggested could be detected with present protocols.

Once the agency has established the relative importance of various possible uses of the data, the rationales for selecting target chemicals should be incorporated into a systematic weighting scheme and applied as comprehensively as is feasible.

Criteria for determining the relative importance of a candidate target chemical should be separated from issues of analytic feasibility until late in the planning. The identification of one or several analytes that might require a new assay protocol could be important in planning future method development.

Problems are likely to occur when method-development projects are concurrent with tissue monitoring (as in the 1982 samples). Design of an adaptable monitoring program with mechanisms for selection of new analytes and for development and validation of collection, storage, and assay methods will permit the monitoring program to remain responsive to current needs and to take advantage of progress in analytic technology.

Formalization of the Planning Process

Present reports do not address the larger issues underlying selection of project goals, nor do they provide insight into the information and alternatives considered in defining those goals. The result is an appearance of arbitrary program decisions; in some cases, decisions regarding the choice of analytic method seem to reflect a "shotgun" approach (e.g., elemental analysis of adipose tissue and possibly the volatile-chemicals projects).

The program would benefit from regular strategic planning by the agency, frequent agency consultation with program contractors, scientific peer review, and advice from interested federal agencies.

Those efforts would maximize the likelihood of a clear decision-making process and one that is well documented and understood.

Critical Evaluation of Program Results

Evaluation of data should be a continuing part of program reporting.

One limitation common to all the project reports produced in the 1980's is that they report very little analysis of final results beyond analytic validity. Data sets are a prime product of a monitoring program, but interpretation of findings in relation to larger program goals (such as time trends, efficacy of interventions, relative importance of different environmental contaminants, and regional or demographic trends in exposures) is an important part of understanding and meeting additional data needs. It seems to the committee that only NHMP itself has broad responsibility for making certain that the program is productive in relation to its larger goals.

Links to Analytic-Methods Research

Analytic-methods research is conducted within EPA, in other government programs, and in academe and the private sector.

The NHATS must be in a position to articulate and, within EPA, influence research priorities for development of new analytic applications of emerging technology and to benefit from new developments.

Analytic-program managers must be specifically charged with the formulation of analytic needs and maintenance of awareness of potentially useful developments. The committee doubts that leadership in analysis can be effectively delegated to contracting organizations, and it believes that EPA must maintain substantially more activity and expertise in this regard.

Regular Schedule of Analyses

The analytic effort has been modified from year to year since 1981, and developmental activities have supplanted monitoring to some degree. The NHATS has released data from only 1 collection year for each of the new sets of analytes. Including current efforts, there are data from only 2 collection years (3 years for the broad-scan pesticides, if the comparability study is included).

Priority should be given to setting and maintaining a schedule for analysis of results of each assay type.

7

Program Design and Management Issues

INTRODUCTION

This chapter deals with the design and management of a new program. Effective management of a newly planned and expanded program for monitoring human tissues requires careful attention to administrative issues. The program must be properly located within the Environmental Protection Agency (EPA), funding must be adequate to accommodate program design and implementation, advisory and peer-review committees must be established for program evaluation and accountability, the professional and administrative capacity must be expanded. For proper operation and implementation of the program, its design must allow changes and flexibility in goals and scope, and some philosophic issues will require specific policy decisions by EPA (e.g., on managing the legal and ethical implications of collecting tissues and on reporting findings of abnormal values).

This chapter concludes with discussion of the importance of timely data analysis and reporting; the importance of cooperation and information transfer with other organizations; timely production and dissemination of reports to enable researchers, policy-makers, etc.; access to and descriptive analysis of the data; and the need for an orderly transition to prevent further deterioration and loss of institutional memory.

ADMINISTRATIVE AND AGENCY ISSUES

Administrative Location

The organizational location of the program is critical. The selection of an agency to spearhead the national monitoring of human tissues for chemical exposures is not simple; indeed, the multiagency history of the NHMP and the current ambivalence within its parent agency regarding its future are clear indications that the match of program goals, potential benefits, and EPA mandates is not perfect. A successful monitoring program must be highly relevant to regulatory needs, but could and should—indeed must—serve a wide range of client programs and must not be dominated by any one of them. Critical aspects of institutional support should not depend on the necessarily changing priorities, requirements, and resources of small subprograms. The NHMP, unlike many other EPA programs, seems to have both a life of its own and a rationale that transcends any of EPA's individual regulatory objectives. And some outputs of a human-tissue monitoring program should be of major interest at the highest levels of the host agency. However, the multipurpose aspect of a national human-tissue monitoring program has meant that, to some extent, the NHMP is not an integrated part of any specific EPA regulatory program and hence is not at the top of any major agenda.

The committee has specific concerns about possible untoward effects of placing a monitoring program in any subunit with direct, major regulatory responsibilities. It firmly recommends that monitoring be kept strictly independent of regulation itself; both fact and appearance are important. Whether the present and past shortcomings in NHMP direction and funding are unavoidable results of a mismatch between agency and program is unclear. The committee concludes that considerations of input to policy, impact, visibility, and independence argue for a location at the highest feasible organizational level. From our consideration of alternative host agencies (based on general concepts of agency mission and resources, with explicit disregard of political and organizational barriers), it is clear that several logical possibilities exist.

Environmental Protection Agency

Advantages of this choice include continuity and a smooth implementation of whatever program changes are required; ease of communication with EPA environmental monitoring efforts; and direct input to a major class of regulatory users of the NHMP products. Such an activity might be placed in the Office of the Assistant Administrator for Research and Development. EPA staff have argued that the NHMP is primarily an environmental monitoring effort and that EPA has both the most appropriate mission and the greatest capability to manage such a program. If a strong and sufficiently independent Bureau of Environmental Statistics (or equivalent) is established, it might also be a suitable home for a monitoring program. Possible disadvantages of EPA as the host agency include its demonstrated difficulty in providing direction and achieving intra-agency and federal support for the NHMP, systematic failure to approach research needs of the program, and failure to "market" results to other users and otherwise to stimulate and support efforts to develop nonregulatory benefits from the program. Other programmatic disadvantages include the lack of an in-house research focus that deals with the biomedical issues of tissue monitoring and a modest (although growing) orientation of present staff to those issues. The ideal host agency must provide technical leadership, as well as funding and administrative direction; that is best achieved with substantial and closely related internal research activity and expertise.

Agency for Toxic Substances and Disease Registry

This alternative would have the advantage that ATSDR is already committed to develop disease and exposure registries, and its mission is more oriented to the biomedical aspects of monitoring than is EPA's. However, ATSDR's activities focus on Superfund concerns, whereas national monitoring will continue to emphasize broad aspects of chemical use and exposure, rather than localized chemical pollution. Comparison data from national surveys are clearly needed by ATSDR for use in interpreting exposures to hazardous waste, but a major effort aimed at national exposure characterization might be too far removed from its mandate. The committee is also concerned that ATSDR is still rather new, growing and evolving rapidly, and perhaps suffering from the "growing pains" inherent in newness and growth.

Centers for Disease Control

Advantages of this choice include strength in epidemiologic research, as well as biomedical and chemical surveillance; demonstrated ability in supporting national monitoring efforts such as NHANES and blood-lead surveys; in-house capability for chemical analyses; and a substantial research focus. Disadvantages include CDC's distance from responsibilities tied to results of monitoring and the fact that CDC is not likely to be a major user of the findings.

Summary

After considering the most likely government units to house a human-tissue monitoring program and hearing testimony at the January 24-25 workshop, the committee recommends that a national program for monitoring human tissues remain within EPA. However, we recommend that its administrative location within EPA be reconsidered and that it be moved to an organizational unit with EPA-wide responsibilities. A location that is geographically close to other programs and laboratories active in the relevant technical disciplines would facilitate important exchanges about methods, as well as followup of findings.

Funding

The critical resources in a program of monitoring human tissues include funding and expertise in appropriate scientific fields.

Sufficient funding (and a reasonable assurance that it will remain adequate over the next few years) is essential. The final budget for a program of this type should be determined after the specifications for the program have been formulated. Major factors that will affect funding requirements are the annual sample size, the set of chemical assays to be performed, the type of tissue to be collected, the size of the in-house staff necessary to monitor the program and analyze results, and requirements for research and development. We do not include in the budget for this program the cost of special studies that might be generated to explore some of the NHMP results in more detail; such studies should, in general, have separately identified funding. Although there is some flexibility in the funding—obviously the greater the funding, the more detailed the analyses that can be carried out—there is a minimum below which it is not worth while to have any program at all. We discuss below alternative budgets that can be considered for this program. The lowest level should be considered an absolute minimum. Furthermore, there needs to be a commitment within EPA to request at least the minimal funding through the indefinite future. The amounts discussed below represent the committee's best judgment of what the various levels could support. These figures are not the committee's judgment of what might be feasible, nor are they recommendations, per se.

An appropriate level of funding for a human-tissue monitoring program is difficult to determine, because it will depend so heavily on the specific program plan developed within the agency. Some guidance is available however, from EPA experience and from similar programs elsewhere. During its public session in January 1989, the committee heard several speakers suggest budgets as high as $50 million per year. Within the United States, a program at NIST for banking liver specimens alone is funded at $150,000. The National Oceanic and Atmospheric Administration National Status and Trends program (consisting of both benthic surveillance and a study of mussels) archives specimens in support of those monitoring programs (human health is not a consideration) and is funded at $100,000 for the banking component only.

Support of the entire EPA National Human Monitoring Program has been $641,000 in FY '87 and $900,000 in FY 1988, FY 1989, and FY 1990. These recent figures for EPA are for external costs only and do not include salaries and overhead of EPA staff members, which in recent years seem to have averaged about two full-time equivalents (EPA, personal communication, June 18, 1990). This level of support at EPA has simply not been enough to sustain a program. The committee has found no one who believes that activity that remains at this level can produce useful results.

The committee firmly concludes that funding should either be increased enough to support a useful program or be eliminated. A clear decision to end the program would be preferable to seriously inadequate support. It would not encourage hopes in the community of present and future users about data that cannot be produced, it would be a clear statement that EPA does not accord human-tissue monitoring a high priority, and it would transfer institutional responsibility out of EPA and perhaps to other federal agencies. However, serious drawbacks to such a decision include the likelihood that no comprehensive, coordinated program would be developed elsewhere, the loss of skilled staff and institutional memory, and perhaps the destruction of specimens that have been banked and saved. The committee sees no likelihood that the information needs documented in Chapter 2 can be filled under this option, unless EPA can find some other willing sponsor. The chance of that seems remote, except possibly for ATSDR, but it is not clear that ATSDR in its present phase of rapid growth, development, and consolidation can spare the top management attention needed for a successful monitoring program. The committee urges that EPA consider termination of human-tissue monitoring only under the most compelling circumstances, and even then only after the fullest exploration of ways to ensure an orderly transfer to some other appropriate agency in less straitened circumstances. Each prior move of the program—from the Public Health Service at CDC to the EPA Office of Pesticides Program (OPP) in 1970 and from OPP to the EPA Office of Toxic Substances (OTS) in 1981 has caused serious disruption, and another move would almost certainly be equally disruptive, even with best efforts to reduce the damage.

It appears to the committee that support at a level of $3 million per year could be barely adequate to sustain the minimal activity needed to keep a program in long-term existence. That is approximately the level of support that was provided in the early 1980s, adjusted for inflation. It was enough then to collect a statistically minimal number of specimens, perform batteries of the most critical chemical tests, maintain the preservation of samples saved from earlier years, and write occasional reports. It was not enough to expand sampling to important geographic areas or population segments not covered, to test for some important chemicals not in the basic panel, to undertake very much in the way of special studies, to prepare records and specimens in a form suitable for use by others, or to undertake the publicity and outreach programs needed to promote widespread use of the collected specimens and data. The committee recommends against a return to the $3-million level, unless careful study shows that circumstances have changed enough for a program of this size to avoid the problems of recent years. That seems unlikely, and the committee is concerned that support at $3 million per year would lead to another decline into unproductive scrabbling and mediocrity, followed in a few years by a new need to decide whether to revive or kill a moribund program. A minimal program might generate some useful data, especially if it were tailored to produce a few important annual reports for a Bureau of Environmental Statistics, or equivalent, but it could not rise to any level of distinction.

Two other critical, specific issues should be addressed by EPA before any decision to implement a minimal program: Is this level of support adequate to attract and maintain the "critical mass" of scientists and technical support staff needed to ensure sampling, testing, and analytic interpretation of the quality needed? Can a program of this size be adequately protected in times of financial exigency, so that it does not again start down the slippery slope on which the effects of decreased resources seem to justify still further decreases? We suspect that the answer to each question is no. Moreover, it would afford no opportunity to collect blood as well as fat specimens. Hence, we do not recommend long-term support at the minimal level, even if it would at first appear to meet the most critical data needs. A larger and broader view is necessary.

The next larger support level the committee considered is $5 million per year, exclusive of staff salaries and overhead. That should be adequate for collection and analysis of at least a statistically representative number of samples per year (see discussion in Chapter 4) plus associated support, research, and analytic activities. We have not undertaken detailed cost analyses, but study of EPA's own history and the operations of other tissue monitoring programs suggests that $5 million per year could support a substantial flow of high-quality, policy-relevant information about chemical burdens in human tissue and that the two special problems just mentioned—maintaining a critical mass of scientific talent and ensuring stability of essential core funding during hard times for the agency—could be solved. It is still not munificent support.

Samples would continue to be small and spotty, so most resulting data would necessarily refer to the whole U.S. population and not more than four to six major geographic segments or population groups; data on say, individual states would not be available, nor would data on small groups of special interest, such as persons living near hazardous waste dumps or working in specific occupational categories (e.g., farmers). Resources for special analyses would be limited, there would be little or no surplus for new chemical tests on stored specimens, and the outreach activities necessary to promote the use of the program outside EPA would be severely constrained. But it might be enough to serve EPA's policy needs and even to bring some critical distinction to the agency. Furthermore, it could be used to develop a solid base of competence, experience, and relevance to promise expansion in the future.

Even greater financial support—even up to the $25-50 million per year suggested by some heads of other agencies (as discussed at the committee's workshop in, January 1989)—could be put to good use, given appropriate planning and the organizational setting and mission described elsewhere in this report. However, such allocations do not seem feasible now, so their implications are not explored here.

Whatever budgetary level is chosen, agency staff, in consultation with the scientific advisory body, should pay continuing attention to the competing demands of adequate sample size (within a context of periodic redesign of the sample to meet changing conditions and needs), protecting and measuring quality at all levels from the selection of subjects to the publication of reports, timely testing for chemicals of current interest, and outreach. A special, critical category will be research within the program—research on improving human-tissue monitoring itself (e.g., on methods of tissue preservation), on improving both the utility and the use of specimens and data, etc. (those matters are discussed in Chapters 4-6).

The committee recommends the prompt allocation of a full-time program manager and other staff with funds adequate for the planning and full design of a new program, concurrent with preparations for the absorption of the NHATS. A support level of at least $3 million per year will be needed for the continuation of current program activities and planning. That part of the program might take 12 months or more after release of this committee report.

We recommend that after the transition period, additional full-time staff be assigned and support be increased to at least $4 million per year for at least 2 years of consolidation. Further growth should be expected, but in a context of competition for funds based on successful implementation of the plan to that time.

Money is clearly essential, but two other kinds of in-house resources are necessary. The first is technical staff with expertise in statistics (primarily survey methods) and in toxicology and related scientific fields. It is acceptable to use an outside contractor for data collection, as now for the NHATS, but the federal agency responsible for the program should be involved in major aspects of the program. It should develop the general plans for data collection, it should monitor the contractor's work, it should plan and oversee data analysis, and it should be prepared to introduce modifications in the program that reflect changes in environmental concerns. That requires staff with deep knowledge of the subject matter and survey methods. (We are not implying that EPA does not have such resources; rather we are describing, in general terms, what an agency needs to carry out a program of the type suggested.)

The second kind of in-house resource that is needed is support at high administrative levels in the agency. That is, of course, necessary to ensure the required funding, but other critical factors are involved. Administrative support informs both the agency staff and the public that the agency has a firm commitment to the program.

The program will require a substantial and continuing research effort in support of improved monitoring, and a large fraction of that should be conducted intramurally or with much more intimate staff involvement than in NHATS contracts. Research activities should be developed in close consultation with other monitoring programs. Examples of subjects that will need research are improved methods for specimen collection and storage, and chemical analytic methods.

Finally, a means should be established to enable the agency to receive advice from outside experts in the pertinent scientific fields. Advice should be sought on such subjects as the performance of the survey (including both the quality of the data and the efficiency of operations), the need to add substances to the analysis because of new environmental concerns, methods of data analysis, and types of reports to be published and their frequency (are discussed below).

Science Advisory Committee

The committee has considered the needs for continuing oversight of a human-tissue monitoring program and concludes that such needs are substantial. The present EPA program has suffered seriously from lack of long-term attention from an outside advisory panel; it might otherwise now be in a far stronger position. Indeed, the committee senses that it is itself serving in some ways as a substitute for oversight that might have been provided in a more timely, effective, and useful way if the highest management levels of EPA had provided sufficient stimulus and internal support.

Such expert advice can be obtained in several ways. One is to have a formal advisory committee on monitoring of human tissue. Another is to engage the service of consultants to make periodic reviews of the progress of the work or of specific issues. A third method is to rely on a broad-based scientific committee advisory to EPA, making sure that the progress of the human-tissue monitoring program is often on its agenda. Other ways are ad hoc reviews by an outside body (such as the present committee), internal program reviews, and continuing attention from some existing advisory body. We are dubious of relying on a general advisory committee; such a committee probably could not devote enough time to a special topic, such as monitoring of human tissues, to make a useful contribution to the program.

The committee concludes that numerous criteria—including cost, breadth and depth of review, objectivity, program stability, timeliness of response, and a continuing need for fresh ideas—point strongly to a standing *outside scientific advisory* body to provide advice and program oversight to management levels of the program and EPA. Each italicized word requires comment.

"Outside" means outside the program and entirely or almost entirely outside EPA, aside from the services of an executive secretary and other support. Staff members from "client" government agencies might be members, as might various persons from academic institutions, industry, and public-interest groups. However, it should be made clear that each member is appointed as an individual and not to represent any organization, bloc, or set of interests.

"Scientific" means that nearly all members should be knowledgeable about one or more scientific and technical disciplines important to the work of the program, such as toxicology, biostatistics, survey statistics, biochemistry, and pathology. Any nonscience members should be selected on the basis of specific, major program needs, such as detailed knowledge about other current programs (surveys of the National Center for Health Statistics come to mind), knowledge about public dissemination of results, or a deep understanding of the policy implications of program findings.

"Advisory" implies that staff should pay substantial attention to recommendations, but make its own evaluations and, for good and clearly stated cause, follow other courses as necessary. The advice should include oversight of the following:

- Program content.
- Program management.
- Program planning.
- Resource needs.
- Technical operations.
- Timeliness and appropriate dissemination of results.

An outside scientific advisory body should also be constituted to facilitate a substantial two-way flow of information and ideas. A role in providing advice to the program is obvious, but role in conveying knowledge about the program and its products back to the range of communities represented on the body is also important.

However the charge is drafted, we believe that the outside scientific advisory body should have no other major responsibilities related to the program. We envision a schedule that initially calls for quarterly meetings, but tapers rapidly to perhaps annual meetings, so that committee assignment will not be seriously burdensome. Members must not, however, be diverted from their primary responsibility for oversight of the human-tissue monitoring program.

The present committee understands the limitations on appointment of new advisory bodies imposed by law, policy, and costs. A standing subcommittee of an existing advisory body might be an appropriate response to the need for outside oversight if most of its members (including the chair) were not members of the parent body.

Other Administrative Issues

Details of program structure and organization will necessarily depend heavily on a host of management decisions that the committee cannot foresee. This section presents some general issues that should be considered and expresses judgments about several broad aspects of structure and organization, but does not reach into details.

An important issue is the role of inhouse competence, both scientific and managerial. The committee recognizes the general competence and expertise of personnel who have been associated with EPA's present program, but sees evidence that such general competence must be backed up within the program itself by substantial first-hand knowledge (training amplified by experience) in each of the technical disciplines relevant to a monitoring program, such as toxicology, biostatistics, pathology, and biochemistry. Contract support, consultants, and borrowed experts can be helpful, but they are not enough. The program requires technically competent staff who are intimately familiar with both the general program needs of EPA (and perhaps other agencies) and emergent problems and who are in close day-to-day contact with potential users of the collected data or stored samples. We are not arguing against appropriate use of contract support, but rather against contracting out so much of the management, technical competence, and programmatic understanding that critical organizational roots are damaged.

A second general issue is the need for some professional staff members to be fully dedicated to human-tissue monitoring. The committee cannot specify full-time scientific disciplines or organizational capabilities, but reaches its conclusion from its perception that EPA's present program has suffered severely from the diversion of time and talent when competing responsibilities have been perceived to be more acute (if not more important). Again, consultants and contract support can be important in program development and support, but they must not be used as substitutes for a core of fully dedicated professionals who have no other, competing duties.

The committee recommends that the program be designed in a modular fashion insofar as is feasible. It is not possible to foretell future strains on resources, trends in the availability of tissues, or needs for program output. It therefore seems prudent to organize a human-tissue monitoring program in such a way that new activities can be smoothly added and so that the most critical core activities can be maintained even when other activities are, for good cause, curtailed, suspended, or even terminated. For example, some panels of chemical tests might be identified as critical for collection every other year, but less critical for annual collection. Oversampling of some population segments might be suspended without reducing efforts to maintain basic coverage of the population as a whole. Preservation of a bank of saved specimens might be assigned higher or lower priority than other important program activities. Judgments about such matters will, of course, be important on initiation of a new monitoring program, but the recommendation here has to do with possible *future* contingencies and the design of activities so that desirable expansion or necessary retrenchment will cause the least possible disruption to core activities. This point could affect a wide range of initial and continuing program decisions, including such matters as the timing of chemical testing of new samples, the storage of preserved specimens (in case some must be discarded), and the design of new series of measurements or reports when projected followup efforts do not come to pass. The committee recognizes the fine line between prudent planning for contingencies and the unintended invitation of inappropriate budgetary cuts, but still recommends that this matter be given some attention from the outset.

It is most important for prudent, frequently up-dated planning to take advantage of resources that become available on short notice. Such planning might best be done, in a widely visible form, in annual budgetary submissions, even when it seems likely that resources will be tightly constrained.

IMPLEMENTATION AND OPERATIONAL ISSUES

Flexibility

Any broad program of human-tissue monitoring should be developed with a view to the need for developmental change, even given the most optimistic expectations regarding the adequacy of initial design. Several factors will make continuing or scheduled periodic redesign essential. As was discussed in some detail in preceding sections covering statistical design, sample collection, chemical analysis, and data analysis, the conditions that govern design choices will inevitably change. For example, demographic and environmental changes can affect the appropriateness of the sampling network, the list of chemical agents included or potentially measurable in the analytic scheme used, the possible choices of tissues based on analytic constraints, and costs. New chemicals will be introduced into use or will be newly recognized as environmental contaminants or risk factors. Advances in analytic technology might permit the recognition of new groups of agents or metabolic transformation products. Reduced assay requirements brought about by improved analytic technologies might permit alternative tissue-collection strategies that are not now feasible. Most monitoring programs can benefit from occasional updating, but the present program will almost certainly require well-managed evolutionary change, because of the open-ended nature of the scientific issues linking environmental contamination with tissue markers of exposure and health effects. Progress in such diverse fields as toxicology, statistical methods for sampling design and data analysis, exposure assessment, and environmental chemistry can be expected to affect the potential uses and value of human-tissue monitoring data.

Although developmental change is essential to the continued value of a long-term human-tissue exposure surveillance program, it is equally important that analytic stability and comparability of data be maintained. Established monitoring procedures (frequency of analysis, sampling design, and assay methods) must not be changed casually. Formal demonstration of comparability of results should precede any alteration in monitoring methods. The use of "probationary" methods in parallel with established methods is one way of managing the implementation of new methods in the monitoring program.

Some strategies for planning for program development are discussed in the preceding chapters. They can be summarized as follows:

• Technical planning needs to be formalized for each year's effort in sample design, sample collection, archiving, chemical assays for monitoring, chemical analysis for program development, and routine and exploratory data analysis. Planning documents that define and resolve design issues should be issued regularly for review by scientific advisers and other agencies.

• Specific budget items to support internal development leading to improved methods and to support external research directed at program needs should be identified.

• Strong external technical input, arising from a scientific advisory panel and based on good communication with related programs and projects, will amplify technical planning by program staff. Equally important is the communication of program needs and findings to outside researchers (whether in other programs or agencies, academe, or the private sector) to stimulate followup.

• Annual reports to Congress and reports for public use should discuss developmental needs and issues.

Legal and Ethical Issues

In addition to the legal and regulatory requirements associated with obtaining human specimens for environmental monitoring, many ethical issues should be addressed. General ethical concerns regarding environmental testing have been outlined elsewhere; some of them depend on the types of tests done to monitor exposures (Belmont Report, 1979; NRC, 1987). One of the important ethical issues to be considered in a monitoring program that uses human tissues is patient confidentiality with respect to the collected and reported data. Patient confidentiality must be safeguarded by handling and analyzing all data on the basis of coded numbers that are used to identify each specimen and related data. The key to identification codes might be broken if additional patient information or samples were needed and if it were clear that this use of identifying information would entail no material risk of further loss of confidentiality.

The codes might also be broken if it were decided to inform patients or their families of extremely high or otherwise atypical concentrations of chemicals detected on routine monitoring. Whether to provide such information is in itself an important ethical issue. In many cases, no therapy will be indicated or even possible for high concentrations of specific chemicals. In others, therapy of family members might be helpful if the conditions of the subjects exposure are such that family members might also have been exposed. Because interpretation of individual measurements of chemicals present in tissue would usually be difficult, a monitoring program must have a clear policy about whether and how such information is to be provided to patients, collectors, or patients' physicians. That issue has been addressed in the NHANES program and has been evaluated by the National Institute for Occupational Safety and Health (NIOSH) and discussed elsewhere (NIOSH, 1988). The NIOSH report demonstrates the difficulty related to disclosure of information under the circumstances of occupational exposures.

The National Institutes of Health is the government agency with primary responsibility for developing and maintaining ethical standards in scientific research that uses humans or human tissues. Living patients must furnish informed consent for the removal of blood or other specimens or tissues primarily for research or monitoring. The consent forms should specify why the samples are being collected, point out that saved (archived) specimens might be used for future projects, and indicate whether results will routinely be provided to the subject, the subject's family, or the family physician. The persons from whom samples are taken should also be informed about the potential complications of specimen collection, including pain associated with obtaining samples (for example, the discomfort of needle biopsies used to obtain samples of fat or other tissue).

Except for blood, most monitoring will use tissues that are *not* removed specifically for research, but are remnants from diagnostic specimens, including autopsies and surgical procedures. Specific patient permission or permission from the next of kin is not required to perform research with such tissues. Such use, along with review of medical records, appears to be in an "exempt" category (OPRR Report 45CFR46, March 8, 1983). Most research institutions include a general permission clause in the surgical or autopsy consent forms to permit use of tissues and nonidentifying patient information in research. Use of such specimens requires strict adherence to both basic tenets of patient confidentiality and specific terms of the informed consent.

Those requirements and guidelines for human research not only apply to research supported by NIH, but also are used generally by institutional review boards (human-use committees) at academic institutions and in industry when such organizations review experimental protocols dealing with humans or with human tissues. In addition, a government committee is preparing uniform requirements for informed consent that will apply to all research supported by any agency of the federal government. European countries tend to be much stricter in their requirements for informed consent from patients whose tissues or medical records are used for medical research (Luepke, 1979).

Some aspects of the collection and use of human tissues for research or monitoring have been complicated by the commercialization of biotechnologic products that were developed from human tissues without specific permission. One commercial use resulted in a major lawsuit [Moore versus the Regents at the University of California et al. (88 C.D.O.S. 5320)] that has been considered by the California supreme court. The fundamental issue in the lawsuit was whether human tissue removed from the body is the property of the person from whom it was obtained. The ruling determined that the tissue did not belong to the patient. Because of this and similar lawsuits, the Office of Technology Assessment recently reviewed the issue of ownership of human tissues (OTA, 1987) and proposed several alternatives for Congress to consider in future legislation. The legal issues associated with use of human tissues in research and transplantation have also been recently reviewed (Swerdlow, 1985).

For the short term, it is important that tissues used for biomedical research or environmental monitoring be separated from those used for the development of commercial products (Grizzle, 1985). Some organizations that supply human tissues for research require that the tissues not be used to make commercial products (Clausen et al., 1989), and a similar proviso should be applied to all tissues collected for environmental monitoring or supplied through the tissue archive of the monitoring program. The issues associated with the commercial use of human tissues are extremely complex and not easily solved, but may in the long run inhibit the availability of human tissues for noncommercial research or environmental monitoring, whether because of donor reluctance to provide tissues that might be used commercially or because of legislation that might inhibit the collection of tissues.

ANALYSIS AND REPORTING OF DATA

Types of Data Analysis and Reporting

The production of data sets that characterize nationally representative average tissue concentrations of specific chemicals is a necessary goal, but not in itself sufficient for a national tissue-based monitoring program. It is highly desirable that the monitoring program develop a plan for making maximal use of raw analytic data to pose and answer questions related to exposures of the U.S. population. However, such an open-ended charge may have adverse consequences; for example, the reporting of results may be delayed while exploratory data analyses are conducted, or the more sophisticated, nonroutine data analyses may be deferred, possibly indefinitely. The plan for each year's monitoring effort should include data-analysis, analytic subprojects (i.e., what questions are to be addressed and what approaches and strategies would be used), and timelines for each activity. Analyses that are well defined should result in early reporting of findings; analyses that require method development or input data beyond the basic chemical-analysis data set might be reported less rapidly or less often. At a minimum, however, descriptive summary analyses of the data should appear with the same frequency as the chemical analyses.

Descriptive analysis should include characterization of central tendency, confidence intervals, and the frequency or prevalence and magnitude of extreme values for the collection set as a whole and for target groups of samples defined by such variables as geography, age, and sex. Results should be compared with those of previous years to identify statistically significant trends. Such comparisons might be difficult or yield uncertain results if sampling methods or analytic procedures change. This most basic ("level 1") data analysis should be presented with information on sample collection, routine chemical assay, and quality control in a report that can form the basis for an annual report of activities. Examples of additional analyses not likely to be completed on the same schedule as level 1 analyses are those that use extramural data (such as toxicologic data or environmental-contamination data) to address questions of dose or routes of exposures and multivariate discriminant functions or factor analyses to assess combinations of results that might be more sensitive than results on individual chemicals ("level 2"). Still more remote in time would be data analyses that require a multiyear data set or input data not currently available ("level 3"). For level 2 analyses, it would be possible to develop explicit plans and schedules, whereas level 3 efforts would depend on the outcome of efforts not part of the NHMP itself and therefore might not be subject to definite schedules. Program planning carried out continuously would include review of data-analysis plans and status and would thereby differentiate realizable goals from goals that are impractical within the constraints of the program.

Integration of Data Sets
with Data from Other Sources

Data from human-tissue monitoring cannot in general be fully interpreted without other information regarding tissue concentrations, patterns of exposures, and the metabolism and toxicology of individual chemicals. Reporting of those should be systematized, but this will require close collaboration of the statistical analysts with persons providing the other information. Comparison data, especially reports of tissue concentrations of chemical agents, should be sought continuously in the scientific literature and in technical reports from complementary programs, such as NHANES. Environmental-survey results produced by other EPA programs will also merit detailed evaluation, particularly if NHMP findings indicate exposures at variance with predictions from environmental data. Information that would be useful in establishing quantitative links between tissue concentrations and exposures is another critical category of comparison data, because risk assessments are generally based on exposure magnitudes. Where data are needed for NHMP purposes and are unavailable, the needs should be identified in annual reports, and other efforts should be made to stimulate research. Strong interagency and intra-agency collaborations in planning and exchange of data will be required for addressing measurement objectives, will facilitate the analysis of monitoring-program results, and will result in a program that is more useful to cooperating agencies.

Reporting Frequency

It is a key requirement that the program produce timely reports on a regular basis. At a minimum, an annual report of level 1 analyses should be produced within a year of completion of the collection of samples. Additional reports addressing more specific questions regarding exposures or trends among chemicals or population groups, or more fundamental issues of exposure assessment, should also appear frequently, but perhaps with a little more lead time.

Review Process for Reports

Draft reports will almost automatically have internal review within contracting organizations, by NHMP program managers, and by EPA quality-assurance coordinators. Additional extra-agency review of draft reports is desirable, especially peer review by individual scientists, by the science advisory group(s) established for the program, and perhaps by other persons with special expertise. Reviewers should be acknowledged in final reports. The program should result in a continual flow of significant findings and summary results suitable for publication in the peer-reviewed scientific literature. Periodic workshops could be convened to obtain wider peer review of the program and its reports.

Scientific reports of findings should be supplemented by documents that establish program goals, plans, and technical strategies and that in other ways summarize the scientific framework of the program. Because application of findings to individual instances of chemical exposure and possible harm is anticipated as an important use, guidelines for interpreting NHMP findings should be formulated. A "user guide" that addresses such issues as representative sampling, population and individual variability, ways of relating dose to tissue concentration and perhaps to health effects, and other concepts necessary for the proper use of the NHMP resource would help to promote the responsible application of results and advise nontechnical users as to the kinds of questions that should and should not be addressed with data from the monitoring program.

Production and Dissemination of Reports

An important role in planning is to lay out a well-defined process for producing a range of outputs with associated responsibilities and a schedule. The schedule should be widely and continually publicized and should be relaxed only under the most compelling circumstances. The full-time scientific staff should bear most of the responsibility for meeting the schedule and should recognize that timely, high-quality reports on important matters are a sine qua non.

The program should have a specific outreach component that is the primary responsibility of at least one person. It is warranted because of the multiuse nature of the program, the wide-ranging interest in the resulting data, and the clear indications that more passive or low-key approaches to publicizing program reports have failed to reach some critical target groups. Technical reports, scientific publications, presentations at scientific meetings, and other normal modes of communication within some segment of the scientific community are necessary but insufficient parts of program "marketing." Additional steps—such as newsletters, preparation of feature articles for nontechnical publications, and presentations to interested nontechnical groups, including legislators—might all be useful. Some care must be taken to define the larger constituency of the program and to update and expand mailing lists or other tools for reaching potential users of findings.

Lines of communication to academic research investigators must be improved. That group not only is likely to be able to put monitoring-program data to important and innovative uses, but also is a likely source of research findings of value to NHMP projects. National research conferences and even a modest extramural-research funding program might be helpful in heightening awareness of the NHMP in universities, colleges, and other research institutions.

A human-tissue monitoring program should be designed as a multiple-user service activity. That creates substantial obligations for assisting users to understand what the program does and does not provide, for timely analysis and publication of results, for specific and helpful guidance in access to archive specimens, and for active marketing of products. The committee recommends that those matters receive high priority in planning and implementation of a new program.

COOPERATION AND INFORMATION TRANSFER WITH
OTHER ORGANIZATIONS

A tissue monitoring and archive program must cooperate and communicate with other branches of EPA, other government agencies, the academic and private sectors, and foreign environmental programs. For example, in the design of the program, the sponsoring organization should use information on tissue collection, shipping, and storage, including proper storage containers. Much information on those topics has been developed for the Federal Republic of Germany (FRG) environmental monitoring program, the archive program at NIST, and the Canadian wildlife-tissue archive. In addition, new information on environmental monitoring is being obtained and evaluated in Sweden and Japan. Those programs might also supply valuable information on analytic methods to aid in the design of a new program in EPA.

Not only are such cooperation and information exchange important in the design of human-tissue monitoring, but continuing information exchange is critical to the efficient operation of the new program. Such communication should not just rely on "published" information, but should take place in part through less formal channels. For example, if the FRG program identifies a chemical in human tissue, rapid information exchange will permit an expeditious search of the chemical in human tissues of the U.S. population. Similarly, the program should be a leader in national and international conferences on tissue monitoring and archiving.

Continuing cooperation and communication of the program with other programs and within EPA and other government agencies are critical. Strong interagency coordination among federal exposure data bases will enhance federal monitoring efforts. For example, if EPA is acquiring air data for a particular location, a corresponding tissue program might be coordinated. If FDA has a market basket pesticide survey for a specified area, the resulting data could be calibrated against tissue concentrations. Other branches of EPA and other federal agencies should use data from the monitoring program, and they should have input without introducing delays as to how the data are collected, analyzed, and reported. They also should have access to the data as rapidly as practicable.

The committee thinks that there might be special value in the joint development of a small set of measurements to be made in similar ways across a broad range of public and private programs or a means of establishing comparability among programs that could lead to a worldwide data base for environmental toxicants that migrate across long distances and across national boundaries. This also might involve a Bureau of Environmental Statistics. Some programs that might be involved in such an effort are the National Contaminant Biomonitoring Program of the Fish and Wildlife Service; the National Pesticides Monitoring Program; various National Oceanographic and Atmospheric Administration programs; and efforts within academia, such as research on organochlorine pesticides in human milk.

TRANSITION

Our main recommendation is that EPA's present program for human-tissue monitoring be phased out as soon as a program based on blood specimens with limited adipose-tissue specimens is implemented. That leaves open the questions of program activities and support levels while the replacement is being developed. We believe that historical continuity can be valuable; that the longer the past series, the greater the value; and that many of the most critical operational changes (e.g., in freezer temperature and in type of storage containers) can be implemented almost overnight. In addition, samples already collected are scheduled for critical analyses that should not be unduly delayed.

Taken together, those points indicate that the present program will be of sufficient value to its successor to merit temporary continued support at a level consistent with its functions of several years ago. The presence of a complete gap of a year or longer—even several months—would risk major damage to public and legislative support, to internal support (if funds, once reallocated, cannot be recaptured), to whatever parts of the collection reporting network are to be preserved, to the maintenance of critical core technical capabilities (including contractors), and to the integrity of data and specimens already in hand. We are concerned, in short, that a too-prompt implementation of the first part of our primary recommendation—the orderly termination of the present program——will create avoidable and expensive impediments to implementation of the second part, a new program. In our view, *the two parts are not separable*, and to attempt to separate them is explicitly contrary to our recommendation.

However, if interim activities should return to the level of several years ago, budgets must keep pace. We recommend continued support at an annual level of at least $3 million until the present program can be dissolved and its talents and resources absorbed into a new program. The budget during the transition period can probably be different from the funding necessary to sustain an operating program. During the first year (or possibly two) of the transition, the main costs will be for planning and testing of the new program and for continuing the NHATS. As the new program is implemented, its costs will become the dominant part of the budget, and the planning costs will shift to oversight of the program and research and development.

SUMMARY AND RECOMMENDATIONS

The committee's primary recommendation is that the present program for human-tissue monitoring be phased out as soon as a blood-specimen program can be implemented.

Design and management of the proposed program will require careful attention to many aspects, including administrative issues, funding, and data analysis and dissemination.

Administrative Location

NHMP's rationale and objectives transcend any of EPA's individual regulatory objectives. Considerations of input to policy, impact, visibility, and independence argue for a location at the highest feasible organizational level.

After considering the most likely government units to house a human-tissue monitoring program, the committee recommends that a national program remain within EPA. Its administrative location within EPA should be reconsidered, and it should be moved to an organizational unit with EPA-wide responsibilities.

The program would benefit from geographic proximity to other relevant programs and laboratories.

Funding

Sufficient funding (and assurance that it will continue) is essential.

Funding either should be increased enough to support a useful program or be eliminated.

Elimination of the program would be preferable to seriously inadequate support. However, such an action would risk the likelihood that no comprehensive, coordinated program would be developed elsewhere, the skilled staff would be lost along with institutional memory, and specimens that have been banked and saved might be destroyed.

The committee recommends the prompt allocation of a full-time program manager and other staff with funds adequate for the planning and full design of a new program.

This should be concurrent with preparations for the absorption of the NHATS. At least $3 million per year will be needed to continue the current program activities and planning.

After the transition period, additional full-time staff should be assigned and support increased.

At least $4 million per year will be needed. Further growth should be expected in a context of competition for funds based on successful implementation. Technical staff with specific expertise will be needed, as will support at high administrative levels in EPA. The program also will require substantial and continuing research efforts in support of improved monitoring.

Scientific Advisory Committee

Numerous criteria, including cost, breadth and depth of review, objectivity, program stability, timeliness of response, and a continuing need for new ideas, point to a standing outside scientific advisory committee.

The body should be outside EPA; nearly all members should be knowledgeable about one or more scientific and technical disciplines important to the program; and the body should be constituted to facilitate a substantial two-way flow of information and ideas. The body should have no other major responsibilities related to the program.

Other Issues

Some professional staff members should be fully dedicated to the program, without competing duties.

The program should be designed in a modular fashion.

This will permit new activities to be added smoothly and core activities to be maintained even if other activities are curtailed.

Analysis and Reporting of Data

At a minimum, descriptive summary analysis of data should appear with the same frequency as chemical analyses. Descriptive analysis should include characterization of central tendency, confidence intervals, and frequency or prevalence and magnitude of extreme values for the collection set as a whole and for target groups of samples defined by variables.

Data reporting should be systematic and conducted with close collaboration between statistical analysts and persons providing other relevant information. Comparison data should be sought continuously from scientific literature and complementary programs.

Continuing cooperation and communication among EPA programs and other agencies are critical.

The program must produce timely reports regularly. At a minimum, an annual report of level 1 analyses should be produced within a year of completion of sample collection.

In addition to internal review, peer review and review of reports by the science advisory group is desirable. Reports should be supplemented by documents that establish program goals, plans, and technical strategies.

References

Allen, B.C., K.S. Crump, and A.M. Shipp. 1988. Correlation between carcinogenic potency of chemicals in animals and humans. Risk Anal. 8:531-544.

Ambe, Y. 1984. The state of the art of the research on environmental specimen banking in Japan. Pp. 33-44 in Environmental Specimen Banking and Monitoring as Related to Banking, R.A. Lewis, N. Stein, and C.W. Lewis, eds. Boston: Martinus Nijhoff.

Anderson, E.L. 1982. Quantitative Methods in Use in the United States to Assess Cancer Risk. Paper presented at the Workshop on Quantitative Estimation of Risk to Human Health from Chemicals, July 12, 1982, Rome, Italy.

Andersson, I., and L. Gustafsson. 1989. Environmental health monitoring system: A research programme based on biological indicators. Ambio 18:244-246.

Autrup, H.B., K.A. Bradley, A.K.M. Shamsuddin, J. Wakhisi, and A. Wasunna. 1983. Detection of putative adduct with fluorescence characteristics identical to 2,3-dihydro-2-(7´-guanyl)-3-hydroxyaflatoxin B_1 in human urine collected in Murang´a district, Kenya. Carcinogenesis 4:1193-1195.

Becker, P.R., S.A. Wise, B.J. Koster, and R. Zeisler. 1988. Alaskan Marine Mammal Tissue Archival Project: A Project Description Including Collection Protocols. NBSIR 88-3750. Gaithersburg, Md.: National Bureau of Standards, U.S. Department of Commerce. 53 pp.

Belmont Report. 1979. National Commission for the Protection of Human Subjects of Biomedical and Behavioral Research. Ethical Principles and Guidelines for the Protection of Human Subjects of Research. OPRR Reports, April 18, 1979. Bethesda, Md.: Office for Protection from Research Risks, National Institutes of Health, U.S. Department of Health and Human Services. 8 pp.

Birnbaum, L.S. 1986. Distribution and excretion of 2,3,7,8-tetrachlorobenzo-p-dioxin in congenic strains of mice which differ at the Ah locus. Drug Metab. Dispos. 14:34-40.

Bryant, M.S., P.L. Skipper, S.R. Tannenbaum, and M. Maclure. 1987. Hemoglobin adducts of 4-aminobiphenyl in smokers and nonsmokers. Cancer Res. 47:602-608.

Buffler, P.A., M. Crane, and M.M. Key. 1985. Possibilities of detecting health effects by studies of populations exposed to chemicals from waste disposal sites. Environ. Health Perspect. 62:423-456.

Bungay, P.M., R.L. Dedrick, and H.B. Matthews. 1979. Pharmacokinetics of halogenated hydrocarbons. Ann. N.Y. Acad. Sci. 320:257-270.

Carrano, A.V., and A.T. Natarajan. 1988. Considerations for population monitoring using cytogenic techniques. Mutat. Res. 204:379-406.

Chopade, H.M., and H.B. Matthews. 1984. Disposition and metabolism of p-nitroaniline in the male F-344 rat. Fundam. Appl. Toxicol. 4:485-493.

Chu, A., and J. Waksberg. 1988. NHANES III Sample Design Final Report. Rockville, Md.: WESTAT.

Clausen, K.P., W.E. Grizzle, V. Livolsi, W.A. Newton, Jr., and R. Aamodt. 1989. The cooperative human tissue network. Cancer 63:1452-1455.

Cochran, W.G. 1977. Sampling Techniques, 3rd ed. New York: John Wiley & Sons.

Comar, C.L., and F. Bronner. 1964. Mineral Metabolism, Vol. 2, Part A. New York: Academic Press.

Conney, A.H. 1982. Induction of microsomal enzymes by foreign chemicals and carcinogenesis by polycyclic aromatic hydrocarbons: G.H.A. Clowes Memorial Lecture. Cancer Res. 42:4875-4917.

Covey, T.R., E.D. Lee, A.P. Bruins, and J.D. Henion. 1986. Liquid chromatography/mass spectrometry. Anal. Chem. 58:1451A-1460A.

Decad, G.M., and M.T. Fields. 1982. Disposition and excretion of chlorendic acid in Fischer 344 rats. J. Toxicol. Environ. Health 9:911-920.

Degkwitz, E., S. Walsch, M. Dubberstein, and J. Winter. 1975. Ascorbic acid and cytochromes. Ann. N.Y. Acad. Sci. 258:201-210.

Dougherty, R.C., M.J. Whitaker, L.M. Smith, D.L. Stalling, and D.W. Kuehl. 1980. Negative chemical ionization studies of human and food chain contamination with zenobiotic chemicals. Environ. Health Perspect. 36:103-118.

Dunn, B.P., and H.F. Stich. 1986. ^{32}P-postlabelling analysis of aromatic DNA adducts in human oral mucosal cells. Carcinogenesis 7:1115-1120.

Elliott, J.E. 1984. Collecting and archiving wildlife specimens in Canada. Pp. 45-66 in Environmental Specimen Banking and Monitoring as Related to

Banking, R.A. Lewis, N. Stein, and C.W. Lewis, eds. Boston: Martinus Nijhoff.

EPA (Environmental Protection Agency). 1976. Toxicology of Metals, Vol. 1. Subcommittee on Toxicology of Metals. Report No. EPA-600/1-76-018. Health Effects Research Laboratory. Research Triangle Park, N.C.: U.S. Environmental Protection Agency.

EPA (Environmental Protection Agency). 1980. Chemical Identified in Human Biological Media, A Data Base. Second Annual Report. Report No. EPA 560/13-80-036A. Washington, D.C.: U.S. Environmental Protection Agency.

EPA (Environmental Protection Agency). 1983. PCBs in Humans Shows Decrease. EPA Environmental News, press release for Monday, May 9, 1983. Washington, D.C.: Office of Public Affairs, U.S. Environmental Protection Agency. 6 pp.

EPA (Environmental Protection Agency). 1986a. Broad Scan Analysis of the FY82 National Human Adipose Tissue Survey Specimens, Vol. I. Executive Summary. Report No. EPA-560/5-86-035. Washington, D.C.: U.S. Environmental Protection Agency.

EPA (Environmental Protection Agency). 1986b. Broad Scan Analysis of the FY82 National Human Adipose Tissue Survey Specimens, Vol. II. Volatile Organic Compounds. Report No. EPA-560/5-86-036. Washington, D.C.: U.S. Environmental Protection Agency.

EPA (Environmental Protection Agency). 1987a. Overview of the National Blood Network. Document No. NBN-SD-01, Final Draft Report. EPA Contract No. 68-02-4243, Task No. 2-44. Office of Toxic Substances, Office of Pesticides and Toxic Substances. Washington, D.C.: U.S. Environmental Protection Agency.

EPA (Environmental Protection Agency). 1987b. National Air Toxics Information Clearinghouse: NATICH Data Base Report on State, Local and EPA (Environmental Protection Agency) Air Toxics Activities, July, 1987. NTIS No. EPA-450/5-87-006. Springfield, Va.: National Technical Information Service. 358 pp.

EPA (Environmental Protection Agency). 1988. Future Risk: Research Strategies for the 1990s. The Research Strategies Committee, Science Advisory Board. Report No. SAB-EC-88-040. Washington, D.C.: U.S. Environmental Protection Agency. 19 pp.

EPA (Environmental Protection Agency). 1990. Protecting the Environment: A Research Strategy for the 1990s. Office of Research and Development. Report No. EPA-600/9-90-022. Washington, D.C.: U.S. Environmental Protection Agency.

Evans, R.W. 1982. Sister chromatid exchanges and disease states in man. Pp. 183-228 in Sister Chromatid Exchange, S. Wolff, ed. New York: John Wiley & Sons.

Everson, R.B., E. Randerath, R.M. Santella, R.C. Cefalo, T.A. Avitts, and K. Randerath. 1986. Detection of smoking-related covalent DNA adducts in human placenta. Science 231:54-57.

Fleiss, J.L. 1981. Statistical Methods for Rates and Proportions, 2nd ed. New York: Wiley. 321 pp.

Flyer, P., K. Rust, and D. Morganstein. 1989. Complex Variance Estimation and Contingency Table Analysis Using Replications. Proceedings of the Survey Research Methods Section, Annual Meeting of the American Statistical Association, August 1989. Alexandria, Va.: American Statistical Association.

Francomano, C.A., and H.H. Kazazian, Jr. 1986. DNA analysis in genetic disorders. Annu. Rev. Med. 37:377-395.

GAO (General Accounting Office). 1988. Report to the Congress: Environmental Protection Agency. Protecting Human Health and the Environment through Improved Management. Washington, D.C.: U.S. General Accounting Office.

Gaylor, D.W., and J.J. Chen. 1986. Relative potency of chemical carcinogens in rodents. Risk Anal. 6:283-290.

Ghanayem, B.I., L.T. Burka, and H.B. Matthews. 1989. Structure-activity relationships for the in vitro hematotoxicity of N-alkoxyacetic acids, the toxic metabolites of glycol ethers.

Gonzalez, J.F., T. Ezzati, J. Lago, and J. Waksberg. 1985. Estimation in the Southwest Component of the Hispanic Health and Nutrition Examination Survey. Proceedings of the Survey Methods, Annual Meeting of the American Statistical Association, August 1985. Alexandria, Va.: American Statistical Association.

Gordon, S.M., L.A. Wallace, E.D. Pellizzari, and H.J. O'Neill. 1988. Human breath measurements in a clean-air chamber to determine half-lives for volatile organic compounds. Atmos. Environ. 22:2165-2170.

Grizzle, W.E. 1985. Commentary. Pp. 37-38 in Matching Needs, Saving Lives: Building a Comprehensive Network for Transplantation and Biomedical Research, J.L. Swerdlow, ed. Washington, D.C.: Annenberg Washington Program.

Gunderson, E.L. 1988. FDA Total Diet Study, April 1982-April 1984, dietary intakes of pesticides, selected elements, and other chemicals. J. Assoc. Off. Anal. Chem. 71:1200-1208.

Hannah, S.A., and L. Rossman. 1982. Monitoring and Analysis of Hazardous Organics in Municipal Waste Water: A Study of Twenty-five Treatment

Plants. NTIS No. EPA-600/D-82-376. Springfield, Va.: National Technical Information Center. 36 pp.

Hansen, M.H., W.N. Hurwitz, and W.G. Madow. 1953. Sample Survey Methods and Theory. New York: John Wiley & Sons.

Harris, C.C., A. Weston, J. Willey, G.E. Trivers, and D. Mann. 1987. Biochemical and molecular epidemiology of human cancer: Indicators of carcinogen exposure, DNA damage and genetic predisposition. Environ. Health Perspect. 75:109-119.

Haugen, A., G. Becher, C. Benestad, K. Vahakangas, G.E. Trivers, M.J. Newman, and C.C. Harris, 1986. Determination of polycyclic aromatic hydrocarbons in the urine, benzo(a)pyrene diol epoxide-DNA adducts in lymphocyte DNA, and antibodies to the adducts in sera from coke oven workers exposed to measured amounts of polycyclic aromatic hydrocarbons in the work atmosphere. Cancer Res. 46:4178-4183.

Hemminki, K.R., E. Grzybowska, M. Chorazy, K. Twardowska-Saucha, J.W. Scroczynski, K.L. Putnam, K. Randerath, D.H. Phillips, A. Hewer, R.M. Santella, T.L. Young, and F.P. Perera. 1990. DNA adducts in humans environmentally exposed to aromatic compounds in an industrial area of Poland. Carcinogenesis 11:1229-1231.

Hemstreet, G.P., P.A. Schulte, K. Ringen, W. Stringer, and E.B. Altekruse. 1988. DNA hyperploidy as a marker of biological response to bladder carcinogen exposure. Int. J. Cancer. 42:817-820.

Hoel, D.G., N.L. Kaplan, and M.W. Anderson. 1983. Implication of nonlinear kinetics on risk estimation on carcinogenesis. Science 219:1032-1037.

Hulka, B.S., and T. Wilcosky. 1988. Biological markers in epidemiologic research. Arch. Environ. Health 43:83-89.

Hunt, W.R., R.B. Faoro, and W. Freas. 1986. Interim Data Base for State and Local Air Toxic Volatile Organic Chemical Measurements. NTIS No. EPA/450/4-86/012. Springfield, Va.: National Technical Information Service. 174 pp.

Ioannou, Y.M., and H.B. Matthews. 1985. p-Phenylenediamine dihydrochloride: Comparative disposition in male and female rats and mice. J. Toxicol. Environ. Health 16:299-313.

Ioannou, Y.M., J.M. Sanders, and H.B. Matthews. 1988. Methyl carbamate: Species-dependent variations in metabolism and clearance in rats and mice. Drug Metab. Dispos. 16:435-440.

Kalinoski, H.T., H.R. Udseth, B.W. Wright, and R.D. Smith. 1986. Supercritical fluid extraction and direct fluid injection mass spectrometry for the determination of trichothecene mycotoxins in wheat samples. Anal. Chem. 58:2421-2425.

Karki, N.T., R. Pokela, L. Nuutinen, and O. Pelkonen. 1987. Aryl hydrocarbon hydroxylase in lymphocytes and lung tissue from lung cancer patients and controls. Int. J. Cancer 39:565-570.

Kayser, D., U.R. Boehringer, and F. Schmidt-Bleek. 1982. The environmental specimen banking project of the Federal Republic of Germany. Environ. Monit. Assess. 1:241-255.

Kimbrough, R.D. 1982. Disposition and body burdens of halogenated aromatic compounds: Possible association with health effects in humans. Drug Metab. Rev. 13:485-497.

Kish, L. 1965. Survey Sampling. New York: John Wiley & Sons.

Kreiss, K., M.M. Zack, R.D. Kimbrough, L.L. Needham, A.L. Smrek, and B.T. Jones. 1981. Cross-sectional study of a community with exceptional exposure to DDT. J. Am. Med. Assoc. 245:1926-1930.

Kutz, F.W., S.C. Strassman, and J.F. Sperling. 1979. Survey of selected organochlorine pesticides in the general population of the United States: Fiscal years 1970-1975. Ann. N.Y. Acad. Sci. 320:60-68.

Kutz, F.W., S.C. Strassman, C.R. Stroup, J.S. Carra, C.C. Leininger, D.L. Watts, and C.M. Sparacino. 1985. The human body burden of mirex in the southeastern United States. J. Toxicol. Environ. Health 15:385-394.

Lewis, R.A., and B. Klein. 1990. A Brief History of Specimen Banking: Storage, Institutions, and Applications. Proc Ecoin Forma: 1989 First International Congress and Exhibition on Environmental Information, Communication and Technology Transfer Bayreuth, Federal Republic of Germany.

Lewis, R.A., and C. W. Lewis. 1979. Terrestrial vertebrate animals as biological monitors of pollution. Pp. 369-391 in Monitoring Environmental Materials and Specimen Banking: Proceedings of the International Workshop, held in Berlin (West), 23-28 October 1978, N.-P. Luepke, ed. Boston: Martinus Nijhoff.

Lewis, R.A., J. Gillett, J.C. Van Loon, J.M. Hushon, J.L. Ludke, and A.P. Watson. 1987. Guidelines for Environmental Specimen Banking with Special Reference to the Federal Republic of German: Ecological and Managerial Aspects. U.S. Man and the Biosphere Program. U.S. MAB Report No. 12. Washington, D.C.: U.S. Department of the Interior, National Park Service. 182 pp.

Lopez-Avila, V., C.L. Haile, P.R. Goddard, L.S. Malone, R.V. Northcutt, D.R. Rose, and R.L. Robson. 1981. Development of methods for the analysis of extractable organic priority pollutants in municipal and industrial wastewater treatment sludges. In Advances in the Identification and Analysis of Organic Pollutants in Water, L.H. Keith, ed. Ann Arbor, Mich.: Ann Arbor Science Press.

Lucier, G.W., and C.L. Thompson. 1987. Issues in biochemical applications to risk assessment: When can lymphocytes be used as surrogate markers? Environ. Health Perspect. 76:187-191.

Luepke, N.-P. 1979. State-of-the-art of biological specimen banking in the Federal Republic of Germany. Pp. 403-409 in Monitoring Environmental Materials and Specimen Banking: Proceedings of the International Workshop, held in Berlin (West), 23-28 October 1978, N.-P. Luepke, ed. Boston: Martinus Nijhoff.

Lutz, R.J., R.L. Dedrick, H.B. Matthews, T.E. Eling, and M.W. Anderson. 1977. A preliminary pharmacokinetic model for several chlorinated biphenyls in the rat. Drug Metab. Dispos. 5:386-396.

Lynn, R.K., C.T. Garvie-Gould, D.F. Milam, K.F. Scott, C.L. Eastman, A.M. Illas, and R.M. Rodgers. 1984. Disposition of the aromatic amine benzidine in the rat: Characterization of mutagenic urinary and biliary metabolites. Toxicol. Appl. Pharmacol. 72:1-14.

Mack, G.A., and J. Stanley. 1984. Program Strategy for The National Human Adipose Tissue Survey. Document No. NHATS-ST-01. Washington, D.C.: U.S. Environmental Protection Agency.

Macleod, K.E., R.C. Hanisch, and R.G. Lewis. 1982. Evaluation of gel permeation chromatography for cleanup of human adipose tissue samples of GC/MS analysis of pesticides and other chemicals. J. Anal. Toxicol. 6:38-40.

Mahaffey, K.R. 1987. Factors influencing biological responses to chemicals. Pp. 315-331 in Mechanisms of Cell Injury: Implications for Human Health, B.A Fowler, ed. New York: John Wiley & Sons.

Mahaffey, K.R., and J.L. Annest. 1986. Association of erythrocyte protoporthyrin with blood lead level and iron status in the Second National Health and Nutrition Examination Survey, 1976-1980. Environ. Res. 41:327-338.

Marshall, W.J. and A.E.M. McLean. 1969. The effect of nutrition and hormonal status on cytochrome P450 and its induction. Biochemical 115:27P (Abstract).

Martin, M.H., and P.J. Coughtrey. 1982. Biological Monitoring of Heavy Metal Pollution. London, New York: Applied Science.

Mathews, J.M. 1988. Absorption, Disposition, Metabolism and Excretion of 1,1,1-Trichloroethane (TCEN). Studies of Chemical Disposition in Mammals, August 15, 1986 - August 24, 1987. Contract No. NO1-ES-65137, Report No. RTI/3662/00-01P. Submitted to the National Institute of Environmental Health Sciences, Research Triangle Park, N.C.

Mathews, J.M. 1990. Metabolism and distribution of bromodichloromethane in rats after single and multiple oral doses. J. Toxicol. Environ. Health 30:15-22.

Matthews, H.B. 1979. Excretion of insecticides. International Encyclopedia of Pharmacology and Therapeutics, Section 113, Differential Toxicities of Insecticides and Halogenated Aromatics, F. Matsumura, ed. New York: Pergamon Press.

Matthews, H.B., J.J. Domanski, and F.E. Guthrie. 1976. Hair and its associated lipids as an excretory pathway for chlorinated hydrocarbons. Xenobiotica 6:425-429.

MRI (Midwest Research Institute). 1988. Quality Assurance Project Plan: Analysis Adipose Tissue for Dioxins and Furans. EPA Contract No. 68-02-4252. Washington, D.C.: U.S. Environmental Protection Agency.

NCHS (National Center for Health Statistics). 1989. Design and Estimation for the National Health Interview Survey, 1985 - 1994, NCHS series 2, No. 110. Washington, D.C.: U.S. Government Printing Office.

NIOSH (National Institute for Occupational Safety and Health). 1988. NIOSH Board of Scientific Counselors Subcommittee on Individual Worker Notification: Meeting Report, June 7-8, 1988. National Institute for Occupational Safety and Health. Washington, D.C.: Department of Health and Human Services.

Nisselson, H. 1987. Evaluation of the Compositing Scheme Used in EED's NHMP. Report prepared for the Office of Toxic Substances, Environmental Protection Agency. Rockville, Md.: WESTAT

NOAA (National Oceanic and Atmospheric Administration). 1988. National Status and Trends Program for Marine Environmental Quality: Progress Report. A Summary of Selected Data on Chemical Contaminants in Sediments Collected During 1984, 1985, 1986, and 1987. NOAA Technical Memorandum NOS OMA 44. National Ocean Service, National Oceanic and Atmospheric Administration. Rockville, Md.: U.S. Department of Commerce.

NOAA (National Oceanic and Atmospheric Administration). 1989. National Status and Trends Mussel Watch Project. Office of Oceanography and Marine Assessment, National Ocean Service, National Oceanic and Atmospheric Administration. Rockville, Md.: U.S. Department of Commerce.

Nomeir, A.A., S. Kato, and H.B. Matthews. 1981. The metabolism and disposition of tris(1,3-dichloro-2-propyl) phosphate (Fyrol FR-2) in the rat. Toxicol. Appl. Pharmacol. 57:401-413.

Nomeir, A.A., Y.M. Ioannou, J.M. Sanders, and H.B. Matthews. 1989. Comparative metabolism and disposition of ethyl carbamate (urethane) in male Fischer 344 rats and male B6C3F1 mice. Toxicol. Appl. Pharmacol. 97:203-215.

Norstrom, R.J., M. Simon, and M.J. Mulvihill. 1986. Gel-permeation - column chromatography clean-up method for the determination of CCDs (chlorinated dibenzodioxins) in animal tissue. Int. J. Environ. Anal. Chem. 23:267-287.

NRC (National Research Council). 1984a. Toxicity Testing: Strategies to Determine Needs and Priorities. Washington, D.C.: National Academy Press. 382 pp.

NRC (National Research Council). 1984b. NHANES. Pp. 31-39 in National Survey Data on Food Consumption: Uses and Recommendations. Washington, D.C.: National Academy Press. 141 pp.

NRC (National Research Council). 1987. Regulating Pesticides in Foods: The Delaney Paradox. Washington, D.C.: National Academy Press. 288 pp.

NRC (National Research Council). 1988. Drinking Water and Health, Vol. 9: Selected Issues in Risk Assessment. Washington, D.C.: National Academy Press. 284 pp.

NRC (National Research Council). 1989. Biologic Markers in Reproductive Toxicology. Washington, D.C.: National Academy Press. 395 pp.

NRC (National Research Council). 1991. Human Exposure Assessment for Airborne Pollutants: Advances and Opportunities. Washington, D.C.: National Academy Press. 321pp.

Nürnberg, H.W. 1984. Section A: Realization of specimen banking. Summary and conclusions. Pp. 23-26 in Environmental Specimen Banking and Monitoring as Related to Banking, R.A. Lewis, N. Stein, and C.W. Lewis, eds. Boston: Martinus Nijhoff.

Osterman-Golkar, S., E. Bailey, P.B. Farmer, S.M. Gorf, and J.H. Lamb. 1984. Monitoring exposure to propylene oxide through the determination of hemoglobin alkylation. Scand. J. Work Environ. Health 10:99-102.

OTA (Office of Technology Assessment). 1987. New Developments in Biotechnology: Ownership of Human Tissues and Cells. Washington, D.C.: U.S. Government Printing Office.

Ozretich, R.J., and W.P. Schroeder. 1986. Determination of selected neutral priority organic pollutants in marine sediment, tissue, and reference materials utilizing bonded-phase sorbents. Anal. Chem. 58:2041-2048.

Perera, F.P. 1987. Molecular cancer epidemiology: A new tool in cancer prevention. J. Natl. Cancer Inst. 78:887-898.

Perera, F.P., and I.B. Weinstein. 1982. Molecular epidemiology and carcinogen-DNA adduct detection: New approaches to studies of human cancer causation. J. Chronic Dis. 35:581-600.

Perera, F.P., K. Hemminki, T.L. Young, D. Brenner, G. Kelly, and R.M. Santella. 1988. Detection of polycyclic aromatic hydrocarbon-DNA adducts in white blood cells of foundry workers. Cancer Res. 48:2288-2291.

Peto, R. 1978. Carcinogenic effects of chronic exposure to very low levels of toxic substances. Environ. Health Perspect. 22:155-159.

Phillips, D.H., K. Hemminki, A. Alhonen, A. Hewer, and P.L. Grover. 1988. Monitoring occupational exposure to carcinogens. Detection by ^{32}P-postlabelling of aromatic DNA adducts in white blood cells from iron foundry workers. Mutat. Res. 204:531-541.

Randerath, K.M., D. Mittal, and E. Randerath. 1988. Monitoring human exposure to carcinogens by ultrasensitive postlabelling assays: Application to unidentified genotoxicants. Pp. 361-367 in International Agency for Research on Cancer. Lyon: International Agency for Research on Cancer.

Reed, D.V. 1985. The FDA surveillance index for pesticides: Establishing food monitoring priorities based on potential health risk. J. Assoc. Off. Anal. Chem. 68:122-124.

Reed, D.V., P. Lombardo, J.R. Wessel, J.A. Burke, and B. McMahon. 1987. The FDA pesticides monitoring program. J. Assoc. Off. Anal. Chem. 70:591-595.

Sabourin, P.J., J.D. Sun, L.S. Birnbaum, G. Lucier, and R.F. Henderson. 1989. Effect of repeated benzene inhalation exposures on subsequent metabolism of benzene. Exp. Pathol. 37:155-157.

Santella, R.M. 1988. Application of new techniques for detection of carcinogen adducts to human population monitoring. Mutat. Res. 205:271-282.

Schafer, L., and K. Overvad. 1990. Subcutaneous adipose-tissue fatty acids and vitamin E in humans. Relation to diet and sampling site. Am. J. Clin. Nutr. 52:486-490.

Schmitt, C.J., M.A. Ribick, l.J. Ludke, and T.W. May. 1983. National Pesticide Monitoring Program: Organochlorine residues in freshwater fish, 1976-79. Resource Publication No. 152. Washington, D.C.: U.S. Fish and Wildlife Service.

Schulte, P.A. 1987. Methodologic issues in the use of biologic markers in epidemiologic research. Am. J. Epidemiol. 126:1006-1016.

Shamsuddin, A.K.M., N.T. Sinopoli, K. Hemminiki, R.R. Boesch, and C.C. Harris. 1985. Detection of benzo(a)pyrene: DNA adducts in human white blood cells. Cancer Res. 45:66-68.

Shum, L.Y., and W.J. Jusko. 1987. Theophylline tissue partitioning and volume of distribution in normal and dietary-induced obese rats. Biopharm. Drug Dispos. 8:353-364.

Smith, R.D., B.W. Wright, and H.R. Udseth. 1986. Capillary supercritical fluid chromatography-mass spectrometry. Pp. 261-293 in Chromatography and Separation Chemistry. Washington, D.C.: American Chemical Society.

Stafford, C.J., W.L. Reichel, D.M. Swineford, R.M. Prouty, and M.L. Gay. 1978. Gas-liquid chromatographic determination of Kepone in field-collected avian tissues and eggs. J. Assoc. Anal. Chemists 61:8-14.

Strassman, S.C., and F.W. Kutz. 1981. Trends of organochlorine pesticide residues in human tissue. Pp. 38-49 in Toxicology of Halogenated Hydrocarbons: Health & Ecological Effects, M.A.Q. Khan and R.H. Stanton, eds. New York: Pergamon Press.

Swerdlow, J.L. 1985. Matching Needs, Saving Lives: Building a Comprehensive Network for Transplantation and Biomedical Research. Washington, D.C.: Annenberg Washington Program.

Taylor, J.K. 1985. Handbook for SRM Users. National Bureau of Standards, U.S. Department of Commerce, Gaithersburg, Md.: National Bureau of Standards, U.S. Department of Commerce.

Tondeur, Y., W.N. Niederhut, J.E. Campana, and S.R. Missler. 1987. A hybrid HRGC/MS/MS method for the characterization of tetrachlorinated-p-dioxins in environmental samples. Biomed. Environ. Mass Spectrom. 14:449-456.

Upton, A.C., T. Kneip, and P. Toniolo. 1989. Public health aspects of toxic chemical disposal sites. Annu. Rev. Public Health 10:1-25.

Vahakangas, K., G. Trivers, A. Haugen, W. Wright, and C. C. Harris. 1985. Detection of benzo(a)pyrene diol-epoxide-DNA adducts by synchronous fluorescence spectrophotometry and ultrasensitive enzymatic radioimmunoassay in coke oven workers. J. Cell. Biochem. Suppl. 9C:1271.

Vainio, H.S., and K. Falck. 1984. Bacterial urinary assay in monitoring exposure to mutagens and carcinogens. Pp. 247-258 in International Agency for Research on Cancer. Lyon: International Agency for Research on Cancer.

Volp, R.F., I.G. Sipes, C. Falcoz, D.E. Carter, and J.F. Gross. 1984. Disposition of 1,2,3-trichloropropane in the Fischer 344 rat: Conventional and physiological pharmacokinetics. Toxicol. Appl. Pharmacol. 75:8-17.

Wallace, L., E. Pellizzari, T. Hartwell, H. Zelon, C. Sparacino, R. Perritt, and R. Whitmore. 1986. Concentrations of 20 volatile organic compounds in the air and drinking water of 350 residents of New Jersey compared with concentrations in their exhaled breath. J. Occup. Med. 28:603-608.

Walles, S.A., S.H. Norppa, S. Osterman-Golkar, and J. Maki-Paakkanen. 1988. Single-strand breaks in DNA of peripheral lymphocytes of styrene-exposed workers. Pp. 223-226 in International Agency for Research on Cancer. Lyon: International Agency for Research on Cancer.

Weinberg, R.A. 1989. Oncogens, antioncogenes, and the molecular bases of multistep carcinogenesis. Cancer Res. 49:33713-33721.

Wise, S.A., and R. Zeisler. 1984. The pilot Environmental Specimen Bank program. Environ. Sci. Technol. 18:302A-307A.

Wise, S.A., and R. Zeisler. 1985. The U.S. pilot Environmental Specimen Bank program. Pp. 34-45 in International Review of Environmental Specimen Banking, S.A. Wise and R. Zeisler, eds. NBS Special Publication 706. Gaithersburg, Md.: National Bureau of Standards, U.S. Department of Commerce.

Wise, S.A., R. Zeisler, and G.M. Goldstein, eds. 1988. Progress in Environmental Specimen Banking. NBS Special Publication 740. Gaithersburg, Md.: National Bureau of Standards, U.S. Department of Commerce. 217 pp.

Woolley, D.E., and G.M. Talens. 1971. Distribution of DDT, DDD, and DDE in tissues of neonatal rats and in milk and other tissues of mother rats chronically exposed to DDT. Toxicol. Appl. Pharmacol. 18:907-916.

Appendices

Appendix A

Workshop Agenda
January 24-25, 1989

Tuesday, January 24th
(NAS Bldg., Lecture Room 125)

8:00 a.m.   Continental breakfast

8:30   Welcome and opening remarks

Session I: NHMP and Its Use/Non-Use

8:45   **EPA Overview of NHMP:** J. Breen, Office of Toxic Substances, US EPA, Washington, DC

9:15   **A User's Perspective:** H. Kang, Director, Virginia Office of Environmental Epidemiology, Washington, DC

9:30   **A Non-User's Perspective:** L. Goldman, Chief, Environmental Epidemiology and Toxicology Section, DHS, Berkeley, CA

9:45   **A General View on Monitoring Programs—**"Toxic Chemicals: Predictive Monitoring and the Nature of Real Systems" R. Lewis, University des Saarlandes, Saarbruecken, West Germany

10:30   Break

10:45   **General Discussion**

Session II: Monitoring Objectives

11:15 **Panel Discussion: "Present and Future Monitoring Objectives—A National Perspective"** V. Houk, CDC, Atlanta, GA; B. Johnson, ATSDR, Atlanta, GA; G. Lucier, NIEHS, RTP, NC; F. Young, FDA, Rockville, MD

12:15 p.m. Lunch

Monitoring Objectives and Specialized Program Needs:

1:15 R. Kimbrough, Director, Risk and Health Capabilities, US EPA, Washington, DC

1:30 S. Sieber, Division Cancer Etiology, National Cancer Institute, Bethesda, MD

1:45 K. Sexton, Director, Office of Health Research, US EPA, Washington, DC

2:00 D. Stevenson, Consultant, Shell Oil Company, Houston, TX

2:15 **Commentator and Formal Discussant:** J. Gannon, US Fish and Wildlife Service, Ann Arbor, MI

3:15 Break

3:30 **Public Session**

4:30 Recess

Wednesday, January 25th

8:00 a.m.   Continental breakfast

8:30   Welcome and opening remarks

Session III: Program Design

8:45   **Collection:** R. Aamodt, Program Director for Cytology/Pathology, NCI, Bethesda, MD

            Storage:
9:00   P. Mazur, Corporate Fellow, ORNL, Oak Ridge, TN

9:15   R. Hay, Head, Cell Culture Department, American Type Culture Collection, Rockville, MD

9:30   **Archiving:** S. Wise, Research Chemist, National Institute of Standards and Technology, Gaithersburg, MD

9:45   **Data Management:** R. Hay, Head, Cell Culture Department, American Type Culture Collection, Rockville, MD

10:00   Break

Session IV: Alternative Monitoring Programs

10:15   **US Fish and Wildlife Service-Great Lakes/Canadian Monitoring:** J. Gannon, US FWS, Ann Arbor, MI

10:30   **National Cancer Institute:** S. Sieber, NCI, Bethesda, MD

10:45   **National Health and Examination Survey (NHANES):** R. Murphy, Director, Division of Health Examination Statistics, Hyattsville, MD

11:00   **Market Basket (Total Diet) Survey:** P. Lombardo, US FDA, Washington, DC

11:15   **Total Exposure Assessment Methodology (TEAM) Study:** L. Wallace, Environmental Scientist, US EPA, Washington, DC

11:30   **Mussel Watch; National Status and Trends Program:** G. Lauenstein, NOAA, Rockville, MD

11:45   **Environmental Specimen Banking Program:** S. Wise, Research Chemist, NIST, Gaithersburg, MD

12:00   **General Discussion**

12:30   Lunch

Session V: Committee Panel Discussion

1:30   **Workshop Discussion:** Committee on National Monitoring of Human Tissues

2:30   **Summary Remarks:** J. Bailar, Committee Chair

2:45   Adjournment

Appendix B

Workshop Participants
January 24-25, 1989

DR. ROGER AAMODT, Division of Cancer Biology and Diagnosis, National Cancer Institute, Bethesda, MD

DR. JOSEPH BREEN, U.S. Environmental Protection Agency, Washington, DC

DR. JOHN E. GANNON, Habitat and Contaminant Assessment Program, U.S. Fish and Wildlife Service, National Fisheries Center-Great Lakes, Ann Arbor, MI

DR. LYNN GOLDMAN, Environmental Epidemiology and Toxicology Section, Department of Health Services, Berkeley, CA

DR. ROBERT J. HAY, Cell Culture Department, American Type Culture Collection, Rockville, MD

DR. VERNON HOUK, Center for Environmental Health and Injury Control, Centers for Disease Control, Atlanta, GA

DR. BARRY JOHNSON, Agency for Toxic Substances and Disease Registry, Atlanta, GA

DR. HAN KANG, Virginia Office of Environmental Epidemiology, Washington, DC

DR. RENATA KIMBROUGH, Risk and Health Capabilities Office of Regional Operations, U.S. Environmental Protection Agency, Washington, DC

DR. GUNNAR LAUENSTEIN, National Oceanographic and Atmospheric Administration, Rockville, MD

DR. ROBERT A. LEWIS, Fachrichtung Biogeographie, Univ. des Saarlandes, Saarbruecken, West Germany

DR. PAT LOMBARDO, Division of Contaminants Chemistry, U.S. Food and Drug Administration, Washington, DC

DR. GEORGE LUCIER, Laboratory of Biochemical Risk Analysis, National Institute of Environmental Health Sciences, Research Triangle Park, NC

DR. PETER MAZUR, Biology Division, Oak Ridge National Laboratory, Oak Ridge, TN

DR. ROBERT MURPHY, Division of Health Examination Statistics, National Center for Health Statistics, Hyattsville, MD

DR. KEN SEXTON, Office of Health Research, U.S. Environmental Protection Agency, Washington, DC

DR. SUSAN SIEBER, Division of Cancer Etiology,National Cancer Institute, National Institutes of Health, Bethesda, MD

DR. DON STEVENSON, Shell Oil Company, Houston, TX

DR. LANCE WALLACE, U.S. Environmental Protection Agency, Washington, DC

DR. STEVEN WISE, National Institute of Standards and Technology, Gaithersburg, MD

DR. FRANK YOUNG, U.S. Food and Drug Administration, Rockville, MD

Appendix C

Summary of Workshop

The workshop conducted by Committee on National Monitoring of Human Tissues was held on January 24-25, 1989, and had several objectives: (1) to gather specific information about EPA's National Human Monitoring Program and specifically, the National Human Adipose Tissue Survey (NHATS), (2) to gather general information about other types of monitoring programs, and (3) to consider the needs for monitoring of human tissues on a national scale.

The workshop was organized to enable committee members to hear presentations from individuals who have used NHATS data or tissues, as well as from individuals who might be likely to use NHATS data (e.g., those involved in public health, regulatory branches of the government, and research organizations).

The committee also invited Dr. Robert Lewis, from the Federal Republic of Germany, to provide a general overview of monitoring and specimen banking from an international perspective (as he is involved with monitoring/ banking programs in Europe, Japan, and the United States). Individuals representing various U.S. government agencies were asked to present national monitoring objectives and to comment on NHATS from their agencies' perspectives (Dr. Vernon Houk, Centers for Disease Control (CDC); Dr. Barry Johnson, Agency for Toxic Substances and Disease Registry (ATSDR); Dr. Fred Shank, Food and Drug Administration (FDA); and Dr. George Lucier, National Institute of Environmental Health Science (NIEHS)).

Several individuals were invited to discuss monitoring objectives with regard to specialized program needs. Others were asked to address specific technical areas that are directly applicable to the NHATS program, including tissue collection, storage and archiving, data management, etc. In addition, several presentations were made about alternative types of monitoring programs narrowly focussed or broadly conceived; these presentations were invaluable to the committee as they defined monitoring and banking objectives, described various methodologies, discussed areas of success and failure, and described budgetary requirements.

MAJOR THEMES

Several themes were repeated during the workshop of both a general and a specific nature.

General Themes

• Human tissue monitoring is valuable and necessary and should be expanded.
• Specimen banking should be an integral part of monitoring programs.
• The goals of monitoring must be clearly defined.
• Monitoring programs are expensive, long term, and often without immediate results; however, funding must be made available as well as institutional commitment.
• Interagency cooperation would increase usefulness of the program and lessen the cost burden to any one agency.

Specific Themes

• The current collection method needs improvement: a close relationship with cooperators should be encouraged; more frequent contact with cooperators is necessary; sterile or clean autopsies should be standard; sterile, insulated containers should be provided for transport to the archive; and proper conditions for storage (before storage in the archive) should be emphasized.
• The storage conditions (at the archive or bank) need to be improved: lower temperatures are necessary (-80°F, or the temperatures of liquid nitrogen); proper storage containers and proper labelling are needed; correct orientation of samples in storage must be maintained. The samples should not be compromised.
• A chain-of-custody record for the specimens is needed with respect to storage conditions at initial collection site, during transportation and at the archive.
• The management of data and their availability need improvement.
• The initiatives taken in improvement of the analytic techniques and in

broadening the range of chemical analyses should continue and be repeated with future samples.

• Quality control of specimens, analytic techniques, and sample stability is necessary, and quality assurance of the data needs to be ensured.

• The statistical design of the program needs to be improved: rural areas need to be represented; samples need to be "representative" of the general U.S. population; more information is needed about the sample in addition to age, sex, and general geographic location, e.g., occupation, area of residence, and diet; a determination of the effects of compositing samples is necessary, as is the development of a standard error around the mean values.

• A number of samples (representative of the population) should be banked specifically for long-term archiving.

• A sufficient amount of material should be banked for future multiple analyses or multiple assays.

• Procedures for sample collection should be standardized, and individuals should be trained in these procedures.

SUMMARY

The workshop began with an overview of the NHMP presented by Dr. Joseph Breen of EPA's Office of Toxic Substances. The NHMP comprises the National Human Adipose Tissue Survey (NHATS), which has collected approximately 20,000 specimens since 1967; the National Blood Network (NBN) which has been funded but not implemented; and the special ad hoc studies that involve human tissue monitoring, e.g., the Veteran's Administration study of Vietnam veterans' exposure to "Agent Orange."

Dr. Han Kang, director of the Office of Environmental Epidemiology described the joint study between the VA and EPA (an NHMP special study). The VA analyzed the adipose tissues of Vietnam veterans collected and archived between 1965 and 1971 by the NHATS program. Dr. Kang found the archive very useful for the analysis of "Agent Orange," which is a mixture of two commercial herbicides, and for the analysis of other chemicals, including dioxin and dibenzofurans.

During the general discussion period, Dr. Vernon Houk criticized the VA study. He commented that the President's Agent Orange Working Council's Science Panel specifically recommended against the use of these adipose tissue specimens for analysis of dioxins.

Dr. Lynn Goldman, chief of the Environmental Epidemiology and Toxicology Section at the California Department of Health Services, had several general comments on human monitoring for contaminants in tissues, as well as several specific suggestions relating to the NHATS program. Dr. Goldman stressed the importance of having data on background levels of contaminants in human tissues and suggested that these data are generally lacking for most xenobiotics other than lead and heavy metals. She found the data collected by EPA's NHMP useful because they were the only data available on background levels of pentachlor, phenol, and dioxin at a time when the California Department of Health Services was conducting studies at Superfund sites contaminated with these compounds.

Dr. Goldman stressed the difficulties that arise when state health department officials are asked to investigate allegations of toxicant exposure of the general public. In most cases, very little reference information is available on toxicant background levels or metabolism. State health department officials, therefore, find it difficult if not impossible, to interpret their findings. Furthermore, the lack of a comparison population often leaves citizens open to anxiety over the findings.

Several suggestions were discussed by Dr. Goldman as methods to improve the NHATS program and to increase its general usefulness:

- Specific studies aimed at contaminant levels and pharmacokinetics in populations that are considered vulnerable, e.g., children and pregnant women.
- The collection of data on background levels and other information that would be useful for assessing the introduction of new technologies.
- A ranking of activities and exposure assessments that are related to environmental clean up and regulation.
- The development of a more complete picture of substances to which the population is being exposed, e.g., broad-scan analysis of chemicals to determine real exposures.
- Screening techniques or bioassays that are less specific but highlight toxicologic end points might make the examination of the population more efficient than the current process.
- An expansion of the exposure data to include rural and urban residences, occupational history, and diet, in addition to age, sex, and geography.
- An increased emphasis on data-base management and encouragement of outside collaboration to ensure inclusion of the data in the scientific literature.
- A closer working relationship with established medical and public-health networks, especially in participating MSAs, and more frequent contact with collectors to encourage a better response rate.

Dr. Goldman concluded her statements by reiterating the need for data on background levels of contaminants of the U.S. population for conducting epidemiological studies, interpretation of findings from epidemiologic studies, planning environmental abatement programs, and exposure assessment.

Dr. Robert A. Lewis, currently working in the Federal Republic of Germany (FRG), gave a broad overview of monitoring and remarked on the status of environmental specimen banking in relation to health and environmental assessment. Dr. Lewis stated that environmental protection should be anticipatory in nature. To translate the principle of anticipatory action into practical policy, a corresponding scientific infrastructure is needed, as well as a comprehensive database to determine the actual status of the environment. The latter requires continuous monitoring of the environment and archiving or specimen banking as a necessary complement to monitoring. The archive can then provide the basis for future retrospective studies.

A panel discussion on national monitoring objectives was held with several individuals from federal agencies, including: Dr. Vernon Houk, director of the Center from Environmental Health and Injury Control at CDC; Dr. Barry Johnson, associate administrator of ATSDR; Dr. George Lucier, chief of the Laboratory of Biochemical Risk Analysis, NIEHS; and Dr. Fred Shank, acting deputy director for the Center for Food Safety and Applied Nutrition at FDA.

Dr. Houk described monitoring objectives and techniques currently employed at CDC and specifically addressed EPA's NHATS. One area of concern at CDC is the analysis of compounds on the ATSDR/EPA list of 200 priority toxicants. Monitoring these compounds to determine exposures is being emphasized, as is the development of methods for analysis of these compounds.

A priority toxicant reference range study is being conducted at CDC and coupled to NHANES III (National Health and Nutrition Examination Survey) in collaboration with ATSDR and the National Center for Health Statistics (NCHS). The focus of this study is to determine background levels of 50 priority toxicants, primarily volatiles, and phenols in 1,000 people with no known excessive exposures. Individuals are being selected by age, sex, and geographic region, a strategy similar to that used in NHATS. Unlike NHATS, however, urban/rural status and individual analysis of samples will be considered. There will be no compositing of samples in NHANES III. Another similar study currently planned with NCHS and coupled to NHANES III will examine other priority toxicants.

In its various exposure/monitoring studies, CDC uses direct measurements of toxicants to assess exposures. According to its collection and storage procedures, a pedigree of the specimen is made, which provides the documentation necessary for future use of the samples. Dr. Houk stated that -70° Fahrenheit or the temperatures of liquid nitrogen are necessary for long-term storage.

Dr. Houk made several comments that were specific to the NHATS:

• The NHATS is useful in detecting trends in levels of toxicants over time. It does not, however, provide population-based data.
• The program complements CDC's efforts in identifying and measuring toxicants that are present in humans at concentrations of concern.
• The design of NHATS and NBN are constrained by practical limitations on sample availability. The data resulting from NHATS reflect substantial numbers of persons but do not allow U.S. population percentile estimates to be made.
• Compositing of samples restricts their usefulness for determining background ranges, i.e., the measured value of a composite sample is the same as the mean of the samples that make up the composite. Thus, the distribution of the results of the composites is a distribution of means rather than a distribution of individual values.

Dr. Houk encouraged EPA to continue its activities, but with added emphasis on population-based estimates of measured toxicants. He further encouraged information exchanges with CDC and EPA laboratories with regard to methods used for chemical analyses of samples. He stressed the need for close attention to accuracy, precision, limit of detection, and quality assurance in test methods. His concluding statement, "[I]f a program is worth doing, it's worth doing right," was echoed many times during the workshop.

Dr. Barry Johnson briefly described ATSDR's mandate and monitoring objectives. He also addressed specific comments to the NHMP/NHATS and made several recommendations for increasing the utility of this program. Dr. Johnson explained that ATSDR was mandated by Congress to implement health-related sections of three laws (Comprehensive Environmental Response, Compensation and Liability Act (CERCLA) 1980, amendments to the Resource Conservation and Recovery Act (RCRA) of 1984, and the Superfund Amendments and Reauthorization Act (SARA) of 1986) that are designed to protect the public from adverse health effects caused by hazardous substances.

The agency has developed ten program areas to aid in the implementation of congressional mandates. Three of the program areas involve evaluation of adverse human health effects and diminished quality of life resulting from exposure to hazardous substances in the environment: health assessments and health studies, toxicological profiles, and exposure and disease registries. Dr. Johnson commented on each of these three areas and explained their relevance to exposure monitoring data. He stated that the presence of population-based exposure monitoring data could be a very useful source of information to ATSDR. He stressed, however, that the agency does not believe that such a national data base of population-based data is available to ATSDR at this time.

Dr. Johnson described specific criteria that ATSDR would consider minimum criteria for any national monitoring/exposure data:

- The data and information must fit the needs of ATSDR (i.e., data must be relevant to ATSDR's charge).
- The data sources must be credible (it should be evident that sound scientific methodology was used to generate those data).
- The data should be readily available, easily interpreted and current.

Dr. Johnson stated that a national chemical exposure monitoring program would likely be used by ATSDR and other agencies if consideration were given to the following objectives:

- The data should reflect the entire U.S. population.
- The program should provide sufficient depth of data to permit subdivisions into key demographic groups.
- The program should provide quality control of its data and quality assurance of its laboratory analyses.
- The program and data should meet the expectations and conditions of scientific peer review.

Dr. Lucier presented a perspective on monitoring objectives that reflect the fact that NIEHS is not a regulatory agency. He stated that NIEHS's primary interest in monitoring programs is in the development and validation of biomarkers. Dr. Lucier also suggested that a national monitoring program should be developed that involves several agencies. The multi-agency approach would enable several organizations with varied expertise to address issues such as funding, sample collection, storage and distribution, priorities for sample access, and ethical issues. Dr. Lucier also concluded with the comment that a program worth doing is worth doing well. He expressed the opinion that well-designed and -conducted national human monitoring program will require more funds than are currently being directed toward the NHMP.

Dr. Shank gave a brief overview of FDA's monitoring programs, including programs that monitor for pesticide residues and toxic elements, including heavy metals, and industrial contaminants. The objectives of the programs vary, but enforcement of pesticide tolerances (established by EPA), and determination of incidence and level of residues in foods are of primary concern.

Continuous studies are directed toward these objectives, e.g., the Total Diet Study, commodity monitoring, and special surveys.

Dr. Shank stressed the need for improved monitoring strategies for the food supply and the environment. He suggested that the nation pursue those programs that provide accurate data on issues most important to public health.

Dr. Renata Kimbrough, director, Risk and Health Capabilities, EPA, raised many questions about the general usefulness of monitoring programs and problems associated with monitoring as well as the expense of monitoring. She raised issues about representative and statistical sampling and commented that it would be very expensive to produce either a representative or statistical sample of the population.

Chlorinated hydrocarbon pesticide levels increase in individuals for various reasons (e.g., advanced age, weight loss, sex differences—levels are higher in men, occupation, diet and individual metabolism). Interpretation of data on pesticide levels, therefore, is very difficult. The problem may be further compounded by the reliability of the analytical methodology employed, the number of laboratories analyzing samples, and reproducibility of the results.

Dr. Kimbrough stated that there are many reasons to continue monitoring programs, even with the difficulties discussed above. The objectives of monitoring should be focused on the development of baseline data, the analysis of trends to determine whether levels of targeted chemicals are decreasing once a regulatory action has been taken, and the development of biological markers as measures of indirect exposure.

Dr. Donald E. Stevenson, chair of the Scientific Committee, American Industrial Health Council, and toxicology consultant to Shell Oil in Houston, contrasted the high priority and strong support of reliable exposure data in the United Kingdom with the type of support this activity has in the United States. Dr. Stevenson believes that, because of the interdisciplinary nature of such activity, there is no one group or agency that has taken "ownership" of all aspects of environmental exposures. Dr. Stevenson further noted the lack of interagency communication with respect to exposure measurements.

Dr. Stevenson stated his bias as strongly in favor of exposure assessment playing a key role in identifying situations which require risk evaluation and in determining public health priorities. Dr. Stevenson stated that exposure assessment should be emphasized in the risk reduction process. In addition, he stated that NHATS has provided a truly unique source of data, and it would be unfortunate if EPA did not continue to analyze a sufficient number of samples to follow continuing trends over the next 5 to 10 years. He emphasized the usefulness of the FY 1982 NHATS study, which provided valuable baseline information. He urged EPA to continue this broader approach.

He stressed the usefulness of the broad-scan approach to determine subsequent sampling strategies and to detect the relatively small number of substances that require more detailed investigation. He also noted that a coordinated interagency strategy is necessary to ensure that all potentially important exposure sources are investigated. This strategy must also be integrated into the national priority strategy for health risk reduction.

Several presentations dealt with specific aspects of the NHATS design. Dr. Roger Aamodt, program director for Cytology/Pathology, National Cancer Institute (NCI), discussed patient rights, quality of tissue, and equitable distribution of tissue as it relates to the NCI program. He stressed the importance of quality control of the tissues and stated that experienced pathologists are called upon to insure appropriate patient diagnosis. He discussed the protocols to collect and process tissue to ensure (1) rapid collection and transfer of tissue from the operating room in fresh condition; (2) accurate pathologic diagnosis by preserving appropriate samples; (3) availability of a pathologist at the time the tissue is collected to enable immediate examination; (4) sample sterility; and (5) important research specimens are not lost by preservation in formalin or in other ways that would destroy their value. Dr. Aamodt reiterated the importance of a close relationship between tissue procurement personnel and the operating room personnel. He also stressed the importance of sterility and freshness of the collected tissue: NCI provides iced, sterile containers for their collectors and instructs them on correct collection and storage procedures. Quality control is very important, and samples are packaged appropriately and shipped in insulated containers in wet or dry ice by overnight delivery service.

The question of biohazard is an important one, i.e., the risk of infection to anyone handling the tissues. NCI will not knowingly collect any tissue from patients who have known transmissible bacterial or viral infections. Furthermore, to ensure the adequacy of investigative procedures, each investigator is required to inform and train all personnel in the dangers of and procedures for handling human tissues. NCI has developed a set of guidelines that provide extensive detailed information and procedures for handling human tissues and body fluids.

Dr. Peter Mazur, a cryobiologist and corporate fellow, Oak Ridge National Laboratory, discussed the problems associated with the freezing of tissues. He stated that freezing is a viable option for storage of adipose tissues, but explained that freezing can either preserve cells or destroy them. Freezing under damaging conditions disrupts cell membranes and internal organelles. Cells that have been damaged by freezing will leak intracellular enzymes when thawed. However, storage temperatures of cells, regardless of freezing conditions, should be the temperature of liquid nitrogen (-196°F) or liquid nitrogen vapor (-150°F); these temperatures will ensure that degradation during storage will not occur. There is strong evidence that at -196°F no thermally driven reactions can occur. In addition, if cell viability is not a matter of concern, then any freezing procedure that gets the tissues rapidly down to -196°F and keeps them at this temperature is acceptable.

Dr. Robert Hay, Cell Culture Department of the American Type Culture Collection (ATCC), discussed storage of cells at the ATCC. This collection of cells is the largest of its kind in the world (2,000 new strains accessioned each year; 40,000 strains are maintained in an inventory of more than 1 million ampules; ATCC distributes 80,000-100,000 cultures a year). Cells at ATTC are stored in liquid nitrogen (-196°F) or in liquid nitrogen vapor, although animal viruses are stored at -70°. Each freezer unit (including alarm system) costs about $13,000. The yearly annual cost for liquid nitrogen is about $1,500/unit. The units are very reliable, and require repair or replacement only approximately every 15 years.

Dr. Steven Wise, research chemist, National Institute of Standards and Technology (NIST), discussed environmental specimen banking and specimen archiving procedures at NIST. Specimens are archived as an alternative way of doing long-term trend monitoring. The advantages of banking specimens are obvious: as analytical techniques change and improve, banked specimens can be analyzed with new techniques. Furthermore, the opportunity exists for doing retrospective analyses on compounds that could not be measured at the time of collection.

Dr. Wise presented a generic design for a tissue archival program based on NIST's experience over the past 10 years. The following considerations were suggested as a necessary part of the specimen banking program:

- A specimen bank should be an integral part of a monitoring program.
- A representative number of specimens should be archived.
- A sufficient amount of materials should be archived for multiple analyses or multiple assays.
- Ample storage space should be available for archived specimens.
- Sample analysis should be based on individual samples rather than on composites, as individual specimens have more meaning than a composite.
- Samples should not be compromised, and procedures for collection and storage should be standardized and well documented.
- Procedures used by all collectors should be the same, and a formalized, detailed protocol for sample collection and storage should be encouraged. If deviations are made from protocols, these should be noted and accompany sample at all times.
- Samples should not be thawed to homogenize them for analyses. Cryo-

genic homogenization procedures (at liquid nitrogen vapor temperatures) are available to prevent thawing of the sample.

• Storage conditions should provide temperatures of at least -80°F, but liquid nitrogen vapor temperatures are preferable.

• Samples should be analyzed for stability, which provides an element of quality control.

Finally, several presentations were made from individuals involved with alternate monitoring programs and program strategies. Dr. John Gannon, U.S. Fish and Wildlife Service (FWS), discussed monitoring and specimen banking in the Great Lakes Region. The program falls under an agreement between the United States and Canada, is administered by the International Joint Commission (IJC), and is funded by the United States and Canada jointly in an effort to monitor toxic chemicals in the Great Lakes.

Dr. Gannon urged EPA and the Committee on National Monitoring of Human Tissues to consider the Canadian Wildlife Service Great lakes specimen banking program as a model. Its capability for retrospective study has been extremely successful. For example, herring gull egg tissue collection and source of dioxin contamination within the Great Lakes was easily determined from tissues banked since 1970. Dr. Gannon encouraged EPA and other federal agencies to establish human monitoring of tissues and to institutionalize specimen banking. He acknowledged, however, the difficulties with securing long-term funding for monitoring and banking programs in a climate of short-term budget cycles.

Dr. Robert Murphy, director of the Division of Health Statistics, National Center for Health Statistics (NCHS), described the NHANES program, which is one of the major data collection systems of NCHS. Dr. Murphy stated that NHANES does have national reference distribution information and a national probability sample. NHANES is a representative sample; the measurements that are taken are those that may have health significance or are descriptive factors that can be useful in looking at health risk factors. NHANES is a very expensive monitoring program (NHANES III will cost $100 million, although in the OMB clearance package it is $115 million).

NCHS encourages the involvement of other public and private agencies. It attempts to ensure continuity of resources with reference to personnel and equipment to run the program. Finally, the resources are available to the program for standardization, quality control, monitoring, criticism, and review of the data for the entire length of the program. Dr. Murphy urged federal agencies and the Committee on National Monitoring of Human Tissues to take into consideration the need for continuity, stability of staff, and collection methods when designing programs that monitor human tissues.

Dr. Pat Lombardo described the U.S. Food and Drug Administration's (FDA) Market Basket (Total Diet) Survey. Dr. Lombardo explained that EPA registers or approves the use of pesticides and established tolerances if use of a pesticide may lead to residues in food, and with the exception of meat and poultry, for which USDA is responsible, the FDA is charged with enforcing tolerances for food shipped via interstate commerce. FDA has carried out a large-scale monitoring program for pesticide residues since the early 1960s. The program has two principal approaches: commodity monitoring to measure residue levels in domestic and imported food to enforce tolerances and other regulatory limits, and the total diet study to measure intakes of pesticides in foods prepared for consumption.

The FDA considers the analytical approach to be critical. Most analyses are carried out using one of five well-tested multi-residue methods that can determine a number of pesticides in a single analysis. Findings from the studies are made public. This program was started in the 1960s and the current version has been in place since 1982. Dr. Lombardo stressed the importance of public perception, and the importance of understandable and timely communication of monitoring data.

Dr. Lance Wallace, environmental scientist, U.S. EPA, described the Total Exposure Assessment Methodology (TEAM) study. The main TEAM study measured the personal exposures of 600 people to toxic or carcinogenic chemicals in air and drinking water. Twenty target chemicals were selected on the basis of their toxicity, carcinogenicity, mutagenicity, production volume, presence in preliminary sampling and pilot studies, and amenability to collection on Tenax. The subjects were selected to represent a total population of 700,000 residents of cities in New Jersey, North Carolina, North Dakota, and California. Each participant carried a personal air sampler throughout a normal 24-hour day, collecting a 12-hour daytime sample and a 12-hour overnight sample. Identical samplers were set up near some participants' homes to measure the ambient air. Each participant also collected two drinking water samples. At the end of the 24 hours, each participant contributed a sample of exhaled breath. All air, water, and breath samples were analyzed for the 20 target chemicals.

Mr. Gunnar Lauenstein discussed the National Oceanic and Atmospheric Administration's (NOAA) National Status and Trends Program (NS&T). This environmental monitoring program is very specific in nature and has well-defined goals, which are to quantify the spacial distribution and long-term temporal trends of contaminants in the marine environment. This goal is accomplished by annual collection of bivalves, bottom feeding fish and surficial sediments around the United States. The Mussel Watch collects specimens from 177 sites around the United States and the benthic surveillance program collects specimens from 50 different sites. NIST is responsible for the organic quality assurance of the analyses; the trace element quality assurance is handled by the National Research Council of Canada. A specimen-banking component of the NS&T program is carried out with the cooperation of NIST.

Dr. Steven Wise discussed the several specimen banking activities of NIST, including the ongoing cooperative efforts with NOAA. Approximately 10 years ago, NIST became involved in specimen banking with EPA to determine the feasibility of long—term storage of different environmental specimens. The EPA Offices of Research and Development, and Health Effects Research funded the pilot environmental specimen bank program. NIST's approach was to gain experience in the various aspects of specimen banking, including collection, storage and analysis. Although the program was initially intended to focus on four areas—human, marine, food and air—funding constraints have limited the focus to human tissues with limited work on the marine sample (e.g., in cooperation with NOAA).

The various programs in which NIST has been associated in cooperation with other agencies are the human liver project with EPA, two NOAA programs and the NS&T Program, two projects with NCI, and the Total Human Diet Program with FDA. The human liver project (collections and banking) was initiated in late 1979. More than 550 specimens have been collected, along with analytical data for trace elements, organic pesticides, and PCBs on about 100 of these samples. In addition, NIST has made a comparison of different storage conditions by storing aliquots of the same samples at varying conditions and evaluating their stability. Furthermore, protocols have been developed for sampling and storage procedures, control samples are banked for quality assurance.

This summary represents the opinions of individual workshop participants. This workshop summary has not been formally reviewed, but has been cleared for transmittal to the sponsor by the NRC Report Review Committee. A formal review will be applied to the committee report to be delivered to EPA in September 1990.

Appendix D

The National Health and Nutrition Examination Survey

"The National Health Survey Act of 1956 authorized the secretary of the Department of Health, Education, and Welfare (now the Department of Health and Human Services), acting through the National Center for Health Statistics (NCHS), to collect statistics on a wide range of health issues. Given this directive, NCHS has conducted health examination surveys for more than 20 years. Among other topics, NCHS collects statistics on determinants of health and on the extent and nature of illness or disability of the U.S. population.

"In 1960-62, the first National Health Examination Survey was conducted. The sample population for this survey was adults 18-74 years old. Two additional surveys were conducted during the 1960s on children 6-11 years old and adolescents 12-17 years old, respectively.

"In 1971, the range of topics included in the survey was extended to include nutritional status. Indexes of nutritional status were to be obtained through a medical history, a dietary interview, a physician's examination, medical procedures and biochemical tests, and body measurements. This survey, the first National Health and Nutrition Examination Survey (NHANES I), was conducted between 1971 and 1974. NHANES II, conducted between 1976 and 1980, extended the age range to include infants 60 months to 1 year old.

"NCHS conducted the Hispanic Health and Nutrition Examination Survey (NHANES, 1982-1984) of persons of Mexican-American, Puerto Rican, and Cuban ancestry residing in the southwestern U.S., the New York City area, and Dade County, Florida. The next national survey, NHANES III and, in the tradition of the past national surveys, continues to be a keystone in providing critical information on the health and nutritional status of the U.S. population.

This information is essential for estimating the prevalence of various diseases and conditions, elucidating mechanisms of disease development, and planning for health policy" (NRC, 1984b).

PURPOSES

"The fundamental purposes of the NHANES are "to develop information on the total prevalence of a disease condition or a physical state; to provide descriptive or normative information; and to provide information on the inter-relationships of health and nutrition variables within the population groups" (R.S. Murphy, Director, Health Statistics Branch, National Center for Health Statistics, DHHS, personal communication). Additional purposes of the NHANES are "to measure the health and nutritional status of the U.S. population and specific subgroups and to monitor changes in health and nutritional status over time.

"Thus, the surveys have provided estimates of the prevalence of characteristics or conditions in the American population, and normative or descriptive data have been developed, such as data on weight and stature. Through successive surveys, repeated collection of these estimates permits the assessment of changes in health and nutritional status over time" (NRC, 1984b).

"The data obtained in NHANES surveys have been extremely important in providing needed information on the prevalence of various health conditions and the distribution of physical, psychological and biochemical characteristics in the U.S. population. However, since many health effects are slow-acting consequences of the long-term exposure to a combination of environmental, dietary, social and demographic factors, the importance of each of these interacting factors can only be known by following the same individuals longitudinally over a period of time. Thus, a followup study of the original adult population of NHANES I, the Epidemiologic Followup Survey, was begun in 1982.

"The primary purpose of the NHANES I Epidemiologic Followup Survey was to investigate relationships between the presence of certain environmental, nutritional, social, psychological and demographic factors and the occurrence of specific diseases. Westat, under contract to the National Center for Health Statistics, attempted to locate and interview 14,407 adults who were age 25-74 at the time they were examined in the original NHANES I survey, which was conducted approximately 10 years earlier. The Followup Survey was funded by the National Institute on Aging (NIA) with additional support from other National Institutes of Health institutes and HHS agencies" (Chu and Waksberg, 1988).

DESIGN

General Features

"One of the purposes of the NHANES II program with respect to nutritional status assessment required that the program continue to use, with some modifications, the same format as NHANES I. To monitor nutritional status, the data collected needed not only to be comparable (at least largely), but also, as in NHANES I, to be collected on a probability sample of the noninstitutionalized civilian population of the United States.

"The general structure of the NHANES II sample design was therefore similar to that of NHANES I. The design was that of stratified, multistage, probability cluster sample of households throughout the United States. The sample selection process involved a number of factors, including the selection of primary sampling units, household clusters, and households. A primary sampling unit (PSU) was a primary location, generally a county of small group of contiguous counties, from which sample housing units and sample persons were selected. When the definition and stratification procedures were completed, 64 PSUs throughout the United States were included in the NHANES II survey plan.

"The clinical examinations and other procedures were conducted in specially designed mobile examination centers. These mobile centers were moved from location to location in a predetermined fashion to achieve economy of operation and to avoid the North in the winter. At any time during the survey period, two mobile examination centers were operating (in different locations) while a third was being relocated. These mobile centers provided a controlled, standardized environment for the clinical examinations and tests. Thus, the clinical procedures could be conducted by a trained staff that moved from site to site with the mobile centers.

"Because of the small number of mobile centers, the logistic constraints involved in moving and setting up the centers, the large number of subjects, and the length of each examination, the total period for data collection was 4 years. The average length of an examination was 2-3 hours, but examinations varied, depending on the age of the subject. For example, the examination of a preschool child lasted no more than 2 hours, and that for an adult no more than 3 hours" (NRC, 1984b).

Statistical Design

"The survey was designed to produce statistics for four broad geographic regions of the Untied States and for the total population by sex, age, race, and income classification. NHANES II was a probability sample of the civilian, noninstitutionalized population of the United States aged 6 months through 74 years.

"For nutritional assessment, three population groups of presume higher risk of malnutrition were of special sampling interest. These groups were pre-school children (6 months to 5 years old), the aged (60-74 years old), and the poor (persons below the poverty levels defined by the U.S. Bureau of Census). These three groups were oversampled to improve the reliability of the statistics generated about them. Although women of child-bearing age were also considered to be at risk of malnutrition, oversampling of them was not necessary, because adequate numbers were included in the sample. A total of 21,000 examined persons was desired as the sample size. The number selected from each of the 64 PSUs was to be between 300 and 600."

"In an initial interview conducted in the household, sociodemographic information and medical histories were collected. A visit to a mobile examination center was then scheduled for each subject. At the mobile center, the physical examination, dietary interview, anthropometry, and other procedures and tests were conducted.

"In the end, 27,801 persons were interviewed and 20,322 persons were examined in the 1976-1980 NHANES (NHANES II). Because not all interviewed persons were examined, appropriate statistical adjustments for nonresponse were made. These adjustments brought the sum of the final sample weights into close alignment with sex, age, and race estimates of the Bureau of the Census at the midpoint of NHANES II" (NRC, 1984b).

METHODOLOGY IN NHANES I, II, AND HHANES

"In all three cycles of HANES (i.e., NHANES I, NHANES II and HHANES) the samples of persons were selected through complex, multistage, highly stratified, cluster samples of households. The selection of sample persons involved the designation of primary sampling units (PSU's), enumeration districts (ED's), households, eligible persons, and finally sample persons (SP's). Each PSU consisted of a county or a small contiguous group of counties; ED;s consisted of households in a small geographic area within a PSU. The target population for NHANES I was the civilian noninstitutionalized population under 74 years of age residing in the coterminous United States, excluding residents of reservation lands set aside for use of American Indians. Alaska and Hawaii were added to the target population of NHANES II. For HHANES the target population consisted of Mexican-Americans, Puerto

Ricans, and Cubans living in selected areas of the U.S. A summary of the designs in all three HANES is given in Table 1-1" (Chu and Waksberg, 1988).

DATA COLLECTION METHODS

"The five major components of NHANES II were a household questionnaire, a medical-history questionnaire, dietary questionnaires, a physical examination by a physician, and special clinical procedures and test, including x rays and test on samples of blood and urine.

"The household questionnaire asked for information on family relationship; some demographic items, such as sex, age, and race of family members; housing; occupation, income and educational level of each family member; and participation in the food stamp program and the school breakfast and lunch program. Two medical-history questionnaires were used; one questionnaire was used for children 6 months to 11 year old, and a different one for persons 12-74 years old. Both household and medical-history interviews were conducted in the respondent's home.

"The dietary interview, physical examination, and special clinical procedures and test (depending on the age of the subject) were conducted when the subject arrived at the mobile examination center. The procedures and tests included body measurements (all subjects); skin-prick tests for allergy (persons 6-74 years old); x rays of cervical spine, lumbar spine(except for women under 50), and chest (persons 25-74 years old except for pregnant women); urine (persons 6-74 years old); and blood (all subjects)" (NRC, 1984b).

QUALITY ASSURANCE

"The quality of the dietary component of the survey was monitored at several levels. All interviews were conducted by dietary interviewers who had at least a bachelor's degree in home economics. Before the survey began, the interviewers were trained both in interview techniques and in coding of food items identified in the 24-hour dietary recall. A manual issued to each interviewer described the procedures to be followed.

"To promote consistency in quality control, dietary interviewers periodically reviewed and evaluated each other's work. At every location, each interviewer tape-recorded two randomly selected subject interviews. These recordings were then evaluated for adherence to procedures.

"Randomly selected 24-hour recall forms were manually reviewed at headquarters before programmatic edits were completed. To detect errors and unusual results form a particular location or from a particular dietary interviewer, the NHANES headquarters staff reviewed the "ranges" of nutrient intake for each respondent" (NRC, 1984b).

DATA PROCESSING

"Preparation of data and reports of the NHANES II findings involved several steps. With some items, such as x-ray pictures, interpretations were required to produce data units that could then be coded. Coded data were key punched into machine-readable form. The data were edited and validated. Sampling weights -- the designated number of people in the population represented by a survey subject -- were determined. For selected measures, imputation procedures for item nonresponse were developed and reviewed. Data were then analyzed and reports were developed" (NRC, 1984b).

DATA REPORTING

"Descriptive, analytic, and methodologic reports were published in *Vital and Health Statistics* (Series 1,2, and 11), a publication of the NCHS. All completely edited, validated, and documented tapes were released for public use through the National Technical Information Service. Information has also been made available in journal articles and in presentations at professional meetings" (NRC, 1984b).

RESEARCH INITIATIVES

"NCHS research is of two kinds: statistical analyses describing the potential contribution of demographic, biologic, and other variables to determinants of health and nutritional status' and development, testing, and transfer of methods involved in conducting surveys and validating data from them. This research is conducted intramurally (that is, within DHHS), often in conjunction with the Food and Drug Administration (FDA), Centers for Disease Control (CDC), or the National Institutes of Health (NIH). No extramural research, as with a grants program, is supported directly by NCHS. Although research in support of the surveys is needed, little funding is available" (NRC, 1984b).

NHANES III

"In preparing for NHANES III, survey planners have solicited suggestions from federal agencies, the Congress, the public-health and nutrition communities, researchers, foundations, and associations. A variety of mechanisms have been considered to gather recommendations from these groups, including letters, meetings, and notices in journals" (Chu and Waksberg, 1988).

MAJOR FACTORS AFFECTING THE DESIGN FOR NHANES III

"A number of important factors had an impact on the development of the sample design for NHANES III. Among these were: (a) the existence of alternative sampling frames and whether they were suitable for NHANES III; (b) the definition of domains of study and corresponding precision requirements; (c) the desired level of effort" (Chu and Waksberg, 1988).

PRECISION SPECIFICATIONS AND DOMAINS FOR ANALYSIS

"There are three basic components of the sample size specifications for NHANES III:

• The first related to the minimum sample size for specified subdomains. This minimum was set at 560 examined SP's for all subdomains. The 560 was established to ensure the desired reliability even for statistics that were subject to relatively high design effects (see table 1-2).
• Superimposed on the cell-size requirement was a provision that the total number of examined SP's in the age-sex groups below were at least as large as the numbers show below. These sample sizes are necessary for the growth-curve analyses planned for NHANES III.
• The third involved the distribution of the total sample size by race-ethnicity. The required sizes (in terms of examined persons) are:

Total	30,000
Black	9,000
Mexican-American	9,000
White and all other	12,000

The set of subdomains for which specified reliability was desired consisted of sex-age groups for the largest race/ethnicity populations in the U.S. The age groups differed among the race/ethnicity domains, and are shown in table 1-3. The subdomains consist of the age groups shown, separately for males and females. There are, thus, 52 subdomains, twice the number of age groups" (Chu and Waksberg, 1988).

Appendix E

Foreign Environmental Monitoring Programs Using Human Tissues or Tissue Specimen Banking

National specimen banking programs are now conducted in Canada, Denmark, Finland, Germany, Sweden, and the United States. A joint specimen banking program by the USA and Canada is now being implemented and the activities of the specimen banking program of Germany are being expanded to include the former German Democratic Republic (R.A. Lewis, personal communication; Lewis and Klein, 1990).

Investigators in Brazil, France, Holland, and Japan have contributed to the development of specimen banking, and Japan has conducted pilot studies on the application of specimen banking to air pollution for more than 7 years (Lewis and Klein, 1990).

GERMANY

History

In 1976, a pilot environmental specimen banking program was begun in the Federal Republic of Germany under the Federal Ministry for Research and Technology. In 1980, an act was passed requiring the assessment of the hazards of xenobiotic chemicals for man and for the environment, and the FRG began to bank human tissues at the University of Münster. In 1981, the Central Bank of the Federal Republic of Germany was inaugurated at the Atomic Energy Center in Jülich. This facility stored in liquid nitrogen vapor human specimens of blood, liver, and adipose tissue as well as samples of other environmental indicators including fish, plants, milk, soil, and earthworms. In 1985, a permanent bank was established by the Ministry for the Environment, Nature Protection, and Reactor Safety.

Current Program

The German program for environmental monitoring and specimen banking has continued to grow and expand and represents the most extensive program of environmental monitoring and environmental specimen banking of human tissues of which we are aware. Collection of human samples is under the control of three institutions, while six institutions (one liquid nitrogen storage facility, a -85°C storage facility, and satellite storage facilities) store human samples for the specimen bank. Six laboratories are responsible for human specimen analysis.

In addition to human samples, the program is collecting and banking a wide variety (approximately 15) of other specimens. There is a fingerprint analysis of each sample before storage. Subsequent analysis depends on specific questions being studied.

Excluding overhead and significant personnel costs, the German program is funded at $2.5-3 million per year.

SWEDEN

A Swedish environmental monitoring program has been in operation since 1978. In 1986, a research program, "Environmental Health Monitoring Based on Biological Indicators" was established, which had as its major objective "to develop methods that will indicate impacts on human health and which can be related to environmental factors" (Andersson and Gustafsson, 1989).

The focus of this program is to evaluate health effects of existing pollution levels, vulnerable human populations to environmental agents, and the synergistic effects of pollutants and other environmental factors. Some specific indicators to aid in this evaluation include DNA-adducts as well as measure of specific health effects such as genotoxicity, reproductive disturbances, neurotoxicity and immunological disturbances such as allergy (Andersson and Gustafsson, 1989). This program recognizes the value of a human biological specimen bank and will evaluate methods to establish such a bank in Sweden. This program will spend 4 million SEK per year through 1991.

CANADA

In Canada, responsibilities and efforts for monitoring the environment are divided between federal and provincial governments. On the federal level, some environmental monitoring and/or specimen banking is performed by